Nudibranchs and related molluscs

Tambja verconis (Polyceridae), Portsea Pier, Vic. Mark Norman

A Museum Victoria
Field Guide to Marine Life

Nudibranchs and related molluscs

Robert Burn

MUSEUMVICTORIA
MUSEUM VICTORIA PUBLISHING

Series editors
Mark D. Norman
Robin Wilson
Melanie Mackenzie

Other titles in the series

Crabs, hermit crabs and allies

Sponges

An introduction to marine life

Barnacles

Shrimps, prawns and lobsters

Published by
Museum Victoria 2015

Museum Victoria
Publishing
GPO Box 666
Melbourne Vic. 3001 Australia
Tel + 61 3 8341 7370 or 8341 7536
publications@museum.vic.gov.au

Museum Victoria
www.museumvictoria.com.au

PRINTED BY
Everbest, China
ORIGINAL DESIGN BY
Propellant
TYPESET BY
the printer's drawers and
Belinda Smullen
Communication Design

National Library of Australia
Cataloguing-in-Publication data:
Burn, Robert.
Nudibranchs and related molluscs / Robert Burn.

Includes bibliographical references and index.
Nudibranchs--Australia, South-eastern.
Other Authors/Contributors:
Museum Victoria.
ISBN: 9780980381382 (pbk.)
9781921833007 (e-pub)
9781921833083 (e-pdf)

593.40994

FRONT COVER IMAGE
Burnaia helicochorda, from south eastern Australia and the North Island of New Zealand, eats small anemones living attached to algae and seagrasses. The genus name honours the writer of this guide; the species name refers to the spiralling digestive gland within the cerata. John Chuk.

BACK COVER IMAGES
Upper: *Sagaminopteron ornatum*. John Chuk.
Second: *Mexichromis macropus*. Leon Altoff.
Third: *Thorunna arbuta*. Platon Vafiadis.
Lower: *Flabellina* sp. RB1. Platon Vafiadis.

Parks Victoria
Healthy Parks Healthy People

Victorian Coastal Council

Port Phillip & Westernport CMA

The Norman Wettenhall Foundation

Funding for the Museum Victoria Field Guides to Marine Life from the Australian Government through a grant from its Natural Heritage Trust is gratefully acknowledged. The grant was facilitated by the Port Phillip and Westernport Catchment Management Authority. Publication of the series is funded in part by the Victorian Coastal Council and Parks Victoria. This title within the series was also supported by a grant from the Norman Wettenhall Foundation.

MUSEUM VICTORIA FIELD GUIDE SERIES TO MARINE LIFE

These field guides to marine life enable the amateur naturalist, beachcomber or environmental scientist to identify the marine animals most commonly found along the shore or in shallow waters along the coast of the state of Victoria, Australia. South-eastern Australia is characterised by a rich marine fauna, with many species found nowhere else. Commonly, species found along the Victorian coast also occur in Tasmania, southern New South Wales, and along the southern coast of the continent through South Australia and into southern Western Australia.

This series aims to cover the common marine animals, and each book deals with a different group. More species live on the Victorian shore and in its shallow waters than are included in each book, and many more inhabit the deeper waters of Bass Strait and beyond. See ***Further Information*** at the end of this guide, and Museum Victoria's website: *www.museumvictoria.com.au*

Museum Victoria encourages individuals to explore the diversity of coastal habitats but discourages unnecessary removal of specimens from their natural environment. Museum scientists are interested in new discoveries, and unusual findings can be reported to the Discovery Centre at Museum Victoria or to Reef Watch Victoria: *info@reefwatchvic.asn.au*

Reef Watch Victoria is a community-based marine monitoring program for Victoria's temperate marine environment. Divers and snorkellers conduct regular surveys at their favourite Victorian reef sites using the Reef Watch monitoring kit. For more information, go to: *www.reefwatchvic.asn.au*

The Marine Research Group (MRG) is a branch of the Field Naturalists Club of Victoria that has had a long and productive working relationship with Museum Victoria, actively undertaking research into Victoria's rich marine life and providing curatorial and survey assistance with Museum Victoria's extensive marine invertebrate collection. The MRG meets regularly, welcomes new members, and can be contacted at: *www.fncv.org.au*

A large proportion of the images and information presented in the MV field guide series has resulted from the 'Under the Lens' and 'Bioscan' partnerships between Museum Victoria and Parks Victoria. Along with ports and harbours, Parks Victoria manages marine national parks and sanctuaries throughout Victoria. For more information go to: *parkweb.vic.gov.au*

CONTENTS

INTRODUCTION

The marine environment of south-eastern Australia is very rich in its diversity of habitats and different kinds of animals. Beachcombers, rock-poolers, divers and photographers always find something of interest, but learning more about their 'finds' often proves far more difficult. This guide is about the nudibranchs and their relatives (otherwise known as 'sea slugs') that can be encountered in the intertidal and divable subtidal waters of the Bass Strait area.

'Nudibranchs and related molluscs' include bubble shells, sea hares, side-gilled slugs, sap-sucking slugs and sea butterflies (pteropods), as well as the true nudibranchs. All have scientific names derived from Latin or Greek; only a very few (mainly tropical) **species** have common names, and these are usually generic or broadly applied rather than given to a single kind of animal. Like their distant relatives, the common garden snails (with external shells) and slugs (without shells), nudibranchs and their relatives vary considerably in form, ranging from those with an external shell, into which the animal can retract for protection, to those with complete absence of a shell (except briefly during embryonic development). Freedom from a protective shell has allowed the enormous elaboration of body form and colouration of nudibranchs, much to the delight of observers, divers and photographers.

Phyllodesmium macphersonae (Facelinidae), Portland Bay, Vic. Leon Altoff

In recent years, regional field guides to the marine biota have touched briefly upon the nudibranch fauna of southern and south-eastern Australia. More detailed guides to the nudibranchs and related molluscs of Western Australia, southern Queensland, and the southern Great Barrier Reef are available, and these and other references are listed in ***Further Information*** at the end of the guide.

Phyllodesmium macphersonae eats the greyish polyps of a stoloniferan soft coral that encrusts the stems of brown algae in the shallows of Western Port, Victoria. Slight differences in colour and body shape between animals throughout its wider distribution in the temperate and tropical Indo-Pacific suggest a complex of **cryptic species** awaiting discovery. The species name honours Hope Macpherson (Mrs Jessie Hope Black), Curator of Molluscs, Museum Victoria, 1947–1963, mentor of the writer's early endeavours in malacology and nudibranch research. This image is of a live 25-mm long specimen from Flinders Pier, Vic., type locality for the species. John Chuk.

What are nudibranchs and related molluscs?

This field guide covers nudibranchs and related marine molluscs, many of great beauty, which until recently have been classified as '**opisthobranchs**' (a technical term meaning the gills are towards the rear). They are part of the **gastropod** (or snails and slugs) class of the **phylum** Mollusca. The many species that completely lack a shell are known far and wide as 'nudibranchs' because of their naked or exposed gills.

Throughout the last few decades, textbook concepts of 'Opisthobranchia' have been challenged by morphology-based and (more recently) molecular studies. It is no longer clear if any precise distinctions can be made between the major groups of opisthobranchs and **pulmonates** (air-breathing slugs), but the classification used throughout this work retains the old groups for consistency with the large body of existing literature. However, the reader should be aware that current research, especially DNA-based analyses, will soon cause scientists to adopt an entirely new classification for nudibranchs and related molluscs as we come to better understand evolutionary relationships. Thus, although much of the literature refers to 'Opisthobranchia' (and that is the concept that also forms the foundation for this field guide), we now know that this assemblage is nothing more than an artificial grouping of gastropod molluscs. The group includes (usually marine) slugs or snails with limpet-like, **bivalved** or bubble shells showing tendencies towards reduction or internalisation of the shell, having a more or less twisted internal anatomy but externally a bilaterally symmetrical body with at least one pair of head **tentacles** or a **headshield**. There are, however, many exceptions to this format, and it should be noted that the true bivalve molluscs are in the separate class Bivalvia. Rather than helping to clarify evolutionary relationships, or even having descriptive value, the 'Opisthobranchia' concept is of historical and—to many of us—emotional attachment. For a current technical discussion of the issues, consult Schrödl *et al.* (2011) in ***Further Information***.

For those using this guide, higher classification and phylogeny are of little consequence; their aim is to identify animals at the genus and the species level, a task rendered endlessly fascinating by the often bizarre shapes and magnificent colouration that make nudibranchs and related molluscs so attractive to rock-poolers and diver-photographers. More than 400 species (different kinds) of these gastropod molluscs are known to occur within the Bass Strait region, at least half of which are found from the intertidal zone down to SCUBA-divable depths.

In form, these animals vary greatly. Some have strong, snail-like shells into which they can withdraw for protection, a few of these even closing the shell opening with a horny **operculum** (trap-door or lid) attached to the rear upper side of the tail. At the other extreme are those with external gill-like structures or appendages (**cerata**) along each dorsal side of the body. Defensive features incorporated into the body surface include nasty-tasting acidic glands just below the skin, and undischarged stinging cells (**nematocysts**) taken up from their food and stored in special receptacles (**cnidosacs**) at the tips of the lateral appendages.

Flabellina sp. RB2 closely resembles other *Flabellina* species occurring in the Bass Strait area, but is distinguished by the rounded, non-tentacle-like corners of the anterior **foot**. *Flabellina* species eat **hydroids**, especially *Eudendrium*, some of which can be seen in the upper left background. 18-mm long specimen, Portsea Pier, Vic. John Chuk.

A number of other marine animals are sometimes called sea slugs, or are mistaken for them. Three common southern Australian gastropods (*Lamellaria ophione, Lamellaria australis* and *Onchidella nigricans*) look very much like oval dorid nudibranchs (nudibranchs of the family Dorididae). All three, however, lack the **rhinophores** and gills that project from the **mantle** of dorid nudibranchs.

Not all 'sea slugs' are nudibranchs

The air-breathing gastropod *Onchidella nigricans* (Pulmonata: Onchidiidae), Flinders, Vic. Chris Rowley.

The gastropod *Lamellaria australis* (Gastropoda: Velutinidae), Flinders Pier, Vic. Glenys Greenwood.

The flatworm *Cycloporus* sp. (Platyhelminthes: Euryleptidae), Portsea Pier, Vic. John Chuk.

The sea cucumber *Australostichopus mollis* (Echinodermata: Holothuroidea: Stichopodidae), Port Phillip Bay, Vic. Mark Norman.

The slug-like or worm-like sea cucumbers (in the tropics, bêche-de-mer), also known as sea slugs, are actually holothuroid echinoderms related to seastars and sea urchins. Polyclad flatworms, too, are sometimes mistaken for nudibranchs, but are usually very soft, very flattened, and underneath lack the very distinct muscular crawling foot.

Concealment

Austraeolis ornata (Facelinidae): adult specimen with conspicuous dark-green and bright-green digestive glands, subtidal, Portsea Pier, Vic. John Chuk.

Many nudibranchs and related molluscs are very good at not being seen. They can be astoundingly similar to the host animal or alga on which they live (and often on which they feed). None dress up the body to hide; instead, most use colour and body shape to blend into the background mosaic for camouflage.

Most bubble shells live within the **substrate** (i.e. are **infaunal**); hence their body colour is whitish and unimportant. The relatively few **epifaunal** (living on the substrate) bubble shells are often extravagantly pigmented animals, seemingly not afraid of predators.

Sea hares and **sacoglossans** (sap-sucking sea slugs) are most often coloured and patterned to match their algal foods. Many sacoglossans are totally green, gaining their pigmentation from the green algae upon which they feed.

Pigmentation of the majority of aeolid nudibranchs is greatly influenced by their hydroid food, the colour of the digestive gland indicating their choice of species. Interestingly, small (<10 mm) subtidal specimens of *Austraeolis ornata* have a pink or red digestive gland, whereas in larger specimens it is always dark greenish brown or black. Undoubtedly, a change in hydroid food occurs at a critical stage of growth. The common *Doto ostenta* also exhibits considerable variation in digestive gland colouration—pink is typical and most common, but sometimes populations are observed in which it is cream, yellowish fawn, deep fawn, brown, dark red, or almost black, suggesting that this species eats a variety of hydroid species.

All species of *Janolus* live upon and feed on erect bryozoans related to *Orthoscuticella*. Pigments from the food are stored within the digestive gland ducts within the cerata, helping create a near cloak of invisibility for the nudibranch upon its **host-food**. In addition, most *Janolus* species have one or more white spots along the middle of the back that, when the animal sits among the fronds of its food, resembles the mouth of an anemone, the surrounding cerata suggesting the ring of anemone tentacles. *Janolus* sp. RB3, which has only been found once. 15-mm long specimen, Portsea Pier, Vic. John Chuk.

The chromodorid nudibranch *Noumea verconis* imitates the colour, pattern and shape of its food—the sponge *Aplysilla rosea* (top). Away from the sponge, the nudibranch stands out as a slow-moving piece of sponge with dorsally projecting rhinophores and gills. Its white egg ribbon will often be found nearby. 40-mm long specimen, Blairgowrie Marina, Vic. John Chuk.

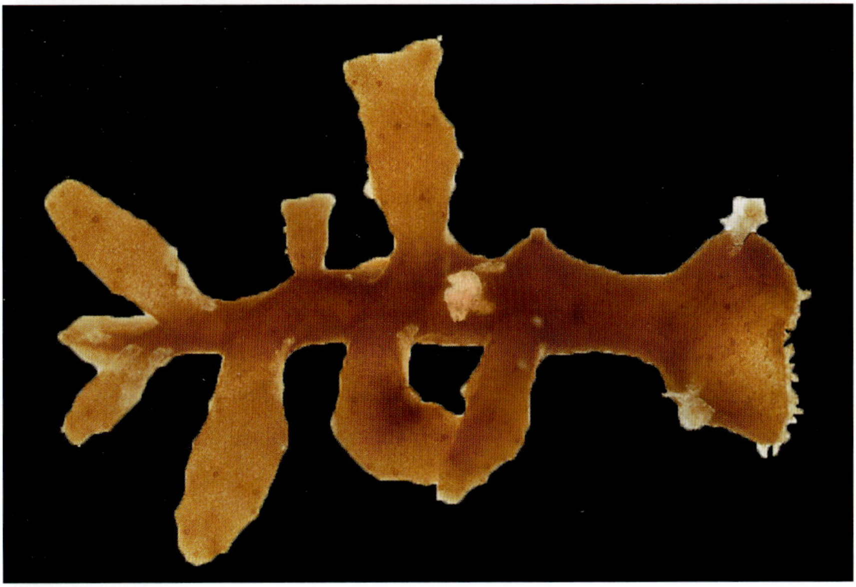

Melibe sp. RB3 is often the greeny brown of larger brown algae, among and over the fronds of which it crawls in its quest for minute crustaceans to capture with its **oral cowl** and then ingest. On brown algae it is difficult to see (top). Off the algae, the body is seen to be angular with a very few large, blade-like cerata along each side. Live specimens of *Melibe* have a sweet smell when handled. 17-mm long specimen, West Head, Flinders, Vic. John Chuk.

Okenia rhinorma ploughs through the sandy bottom searching for its food, the sand **ascidian** *Molgula*. Once food is found, the nudibranch dives under the sand surface, eats a hole through the ascidian skin, and proceeds to **eviscerate** its food. The series of images show the nudibranch ploughing along, and then starting to burrow down to attack its food. 18-mm long specimen, Blairgowrie Marina, Vic. John Chuk.

Biotic associations

Some nudibranchs have remarkable associations with their food organisms and are able to retain functioning elements of their food within their own body.

Plastids are structures within algal cells responsible for photosynthesis. These are found in very many of the sap-sucking sacoglossan nudibranchs. The ingested plastids are retained within the ducts of the digestive gland within the body and the cerata, where they are often clustered as small darker points at the near-surface tips of the fine **ductules**. Good examples are found among the **aeolidiform** or cerata-bearing species *Stiliger smaragdinus*, *Polybranchia pallens* and *Hermaea evelinemarcusae*.

With one or two exceptions, all sacoglossan nudibranchs cut a minute slit in the body wall of green (mostly) or red algae and suck out the algal sap. These images are of the small slender species *Hermaea* sp. RB1 (top) and *Hermaea evelinemarcusae* (below), the former hunched up and resembling a small distasteful anemone with a whitish or pinkish mouth pore, the latter showing the algal plastids responsible for the pigmentation of the digestive ducts within the body and cerata. Different forms of digestive gland branching within the cerata help differentiate species. Both less than 10 mm long, Portsea Pier, Vic. John Chuk.

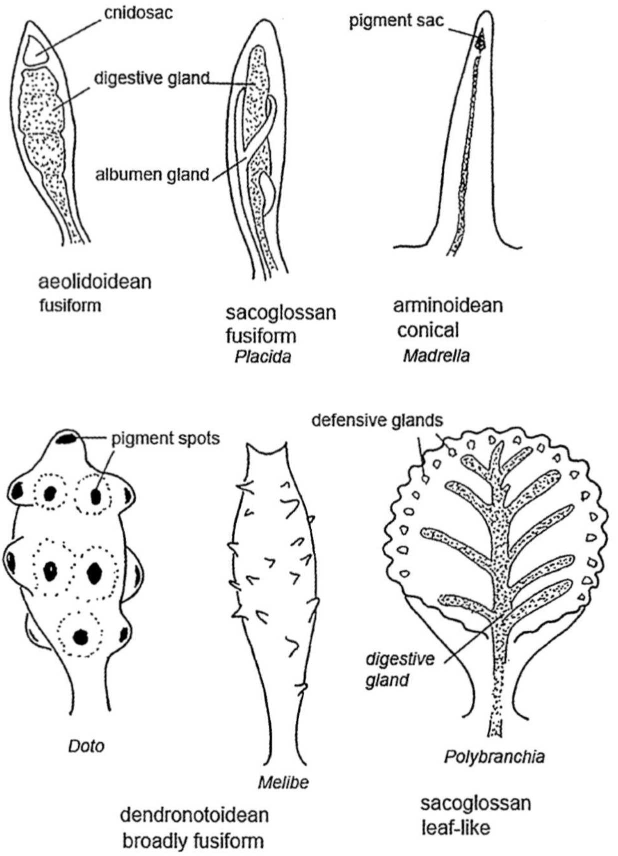

Various forms of cerata, digestive glands and associated structures in nudibranchs and related molluscs. Illustration: Robert Burn, originally published in Burn (1989).

Nematocysts are the stinging cells of **cnidarians**. Somehow or other, many (especially aeolid) nudibranchs are able to ingest, process and store nematocysts in special receptacles, the cnidosacs, in the tip of their cerata, all without causing the nematocysts to explode. It is assumed that the stored nematocysts are used for the nudibranch's defence. Among the Bass Strait aeolid fauna, only *Phyllodesmium* lacks nematocysts.

Burnaia helicochorda (Aeolidiidae), one of many aeolid nudibranchs that retain cnidarian nematocysts in their cerata, Kitty Miller Bay, Phillip Island, Vic. Leon Altoff.

Zooxanthellae are single-celled, photosynthesising dinoflagellates, which can be free-living or found within corals, and with which they have their own **symbiotic** relationship. *Phyllodesmium* gains zooxanthellae from its soft coral prey and stores and cultivates these in the digestive ducts of the body and cerata. *Phyllodesmium macphersonae,* pictured in the introduction to this guide, is an excellent example, the fine ducts of the digestive gland all filled with grey zooxanthellae.

External and internal anatomy

Within the nudibranchs and related molluscs, each of the groups within the traditional classification has diagnostic characteristics to set it apart, but sometimes distinctions are not obvious without careful observation and microscopic study of the various parts of the body.

As in the vast majority of gastropods, the muscular pharyngeal part of the digestive system contains jaws and a **radula**, though exceptions occur where one or other or both are absent. Jaws are used for grasping food, the radula for rasping tissues from the food and passing them rearwards into the oesophagus. The radula is comprised of a number of transverse rows of chitinous teeth attached to a basal membrane. During the course of growth, old worn teeth are continually replaced by new rows of sharp teeth. Old teeth drop off the basal membrane, pass through the gut and are expelled in the faeces. Only in the sap-sucking Sacoglossa and a few of the Aeolidina are a lifetime of teeth retained within the **pharynx**.

Section of radula from *Marianina rosea* showing central tooth **denticulate** on each side of the cusp, the lateral tooth with cusp and **denticles** along the outer edge, and the 4–5 long, smooth marginal teeth to complete the half row.
Illustration of scanning electron microscope (SEM) image: G. Avern, from Beesley, Ross and Wells (1998).

Radular teeth are often very distinctively shaped, which can be useful in classification at genus and family levels, but rarely to the degree that allows species identification. Of importance when examining an animal's radula are the number of transverse rows of teeth, the number of teeth per transverse row, the presence or not of a central (or **rhachidian**) **tooth**, and differentiation in size and shape of teeth along the half transverse row. Radular studies can only be accomplished with a very good optical microscope or **SEM**.

A few small groups of nudibranchs and related species lack radular teeth. These species invariably suck in their food, which is then passed to the gizzard and stomach where enzymes break it down for absorption (Retusidae, Aglajidae). Some dorid nudibranchs (Dendrodorididae) extrude enzymes from glands that open into the mouth and externally digest their sponge food into a soup, which the animal then sucks up.

Many cephalaspidean bubble shells, all sea hares and some dendronotine nudibranchs have, in addition to jaws and radula, hard crushing, abrading or sorting plates or ridges in the gizzard or stomach. These are often distinctive in size and shape, and are useful in separating genera and species.

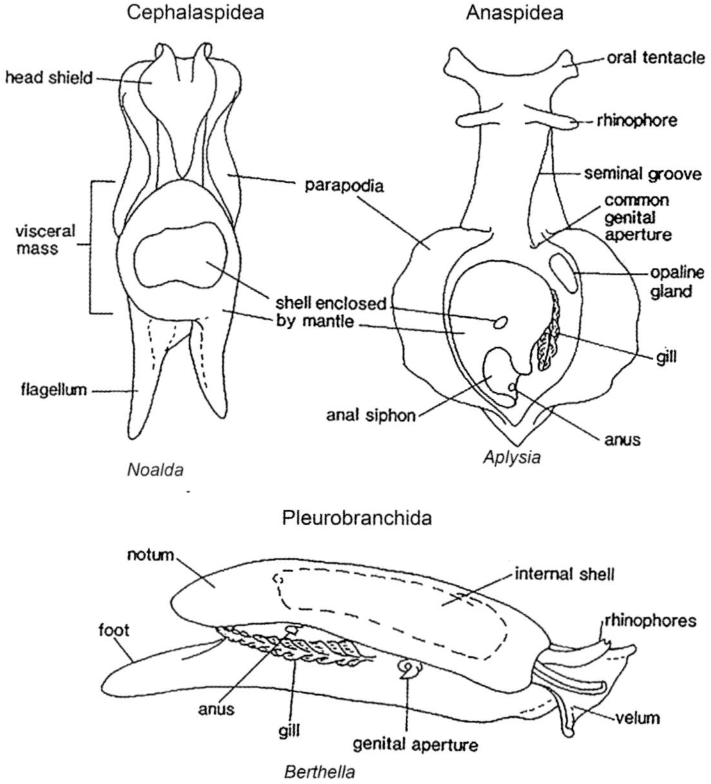

Anatomy of bubble shells, sea hares and related molluscs.
Illustration: Robert Burn, originally published in Burn (1989).

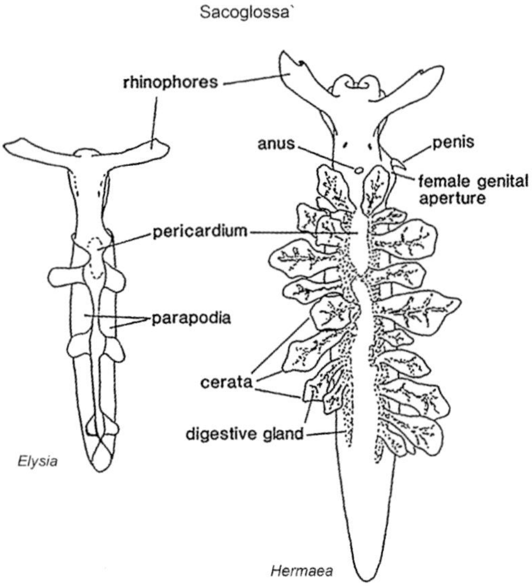

Anatomy of sacoglossans. Illustration: Robert Burn, originally published in Burn (1989).

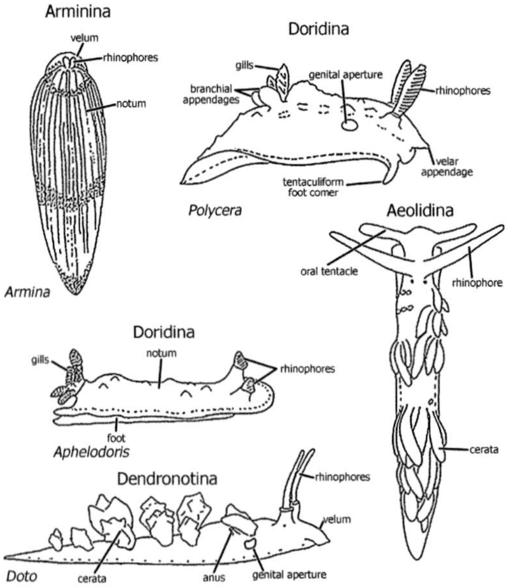

Anatomy of dorid and aeolid nudibranchs and related molluscs. Illustration: Robert Burn, originally published in Burn (1989).

All nudibranchs and related molluscs have **chemosensory** organs that help guide them to food or mates. All (except the bubble shells) have a pair of erect tentacles projecting from the head part of the body. These are the rhinophores. Rhinophores may be smooth and tapering, **papillate** all round or only behind, **annulate** with two to many ring-like projections along the length, or **lamellate** with few to many projecting leaves more or less horizontal, angled or vertical. The chemosensory organs in bubble shells, otherwise known as **Hancock's organs**, can be seen as an often yellowish or brownish patch or ridge on the body wall either side of the mouth. In the Pleurobranchida, the rhinophores project from under the anterior mantle edge, in the Umbraculida from under the shell.

With the exception of the sap-sucking Sacoglossa, the nudibranchs and related molluscs all have **oral tentacles**, lobes or a **velum** close by or above the mouth. These help orientate the animal onto its food. In bubble shells, the lobes are small, raised pads close to the mouth. One very small species (*Cylichnatys campanula*) has lobes that are ruby red in colour, and in the family Aglajidae, **hyaline** sensory bristles may project. Most nudibranchs have oral tentacles, but in dorids these are invariably hidden below the mantle brim. Species of Pleurobranchida, Dendronotina and Arminina (and a very few species of Aeolidina) have a broad velum at the front of the animal, below which is the mouth. The corners of the velum may project as short tentacles, sometimes with a lateral groove. In Dendronotina, the velum is usually fringed anteriorly with a number of long projecting points, in Dotidae, the velum is smooth and, in Tethydidae, it is exceptionally developed for use as a 'casting net' to capture minute crustaceans for food.

The nudibranch gill is most obvious in dorid nudibranchs, projecting like a bunch of branching feathers from the rear half of the mantle. In the orders Umbraculida and Pleurobranchida, the gill lies along the right side of the body wall between the mantle and foot. In some nudibranchs, gill plates or ridges occur along each side of the body below the mantle. In bubble shells and sea hares, the gill is towards the posterior end of the body, within the mantle cavity under the protection of the shell. Shell-bearing species of the order Sacoglossa also have a gill within the mantle cavity, but the shell-less species, as well as many nudibranchs, all achieve gaseous exchange through the epithelium of the body and projecting appendages and cerata.

Reproduction

With only one or two known exceptions, nudibranchs and related molluscs are **hermaphroditic**. In other words, these so-called lowly or primitive animals have a reproductive system that is complete and successful as both male and female. Both male and female **apertures** are on the right side of the body, and sometimes, especially among nudibranchs, both apertures open within a common chamber called an atrium. In other groups, such as bubble shells and sea hares, they are far apart, with the female aperture in the posterior mantle cavity and an open sperm groove running forward to the male aperture, which is almost beside the mouth.

To mate, a pair of animals come together anterior right side to anterior right side so that the reproductive apertures are aligned. Each everts its penis to lock into the keyway that is the female aperture of the partner. Sperm is passed from one to the other and stored for subsequent fertilisation of the eggs as they are produced prior to laying. Depending upon the species, mating may be very brief (only a few seconds in those that are very small) or continue for hours in some of the larger species. Pairs of bubble shells and sea hares mate as above, but when numerous animals are present at the height of the reproductive season, specimens of some species form short chains, with the animal in front being female, the one behind male. Other species gather to form frothy orgiastic groups.

A few very small, shell-less species are known to produce **spermatophores**, which they affix to a partner. The packet is then absorbed into the body, and the sperm finds its way to the appropriate storage area. The outer parts of the male duct may be distinctively shaped, or armed with an apical style, hooks, scales, pads, ridges or protuberances. In rare cases, the female duct is armed with scales.

Nudibranchs and related molluscs all lay eggs and encase them in a gelatinous string, ribbon or balloon. Some species produce a reduced number of relatively large eggs, from which crawling young emerge. Many species produce innumerable very small eggs, which hatch as minute **veliger** ('velum-bearing') larvae that swim in the water column until such time as they are stimulated by age or the presence of the right food to settle and metamorphose into juveniles of the adult species. There are many other intermediate forms of development. An embryonic shell is present in all species, sometimes only briefly within the egg capsule, but in most species the shell is cast aside at metamorphosis.

Philine angasi (Philinidae) egg balloon, Kitty Miller Bay, Phillip Island, Vic. Platon Vafiadis.

Tangled egg mass of a species of *Aplysia*, Portsea Pier, Vic. Mary Molloy.

The egg ribbons, strings or balloons of nudibranchs and related molluscs are sometimes seen before the laying or depositing animal is noticed. Bubble shells lay their eggs within gelatinous balloons anchored by a short, stout stem and holdfast to sand grains or a shell. Those of *Philine angasi* may be 75–145 mm long by 20–25 mm in diameter, and they are often found on the shallow, sandy shores of protected bays and estuaries. Other bubble shells lay minute, spherical balloons attached to sand grains or algae, which can make them difficult to observe. Sea hares, especially of the genus *Aplysia*, produce tangled masses of string-like egg ribbons, often cream when first laid, but darkening to other colours before hatching.

Nudibranchs, pleurobranchs and umbrella shells lay spirally coiled egg ribbons on hard surfaces. Some ribbons are flat, others are erect, often with the upper free edge waving about in the current. The young either hatch from the eggs as free-swimming veligers that live in the water column until such time as they settle and metamorphose into miniature adults, or develop within the egg capsule and emerge as crawling young. Adult animals die after completing the egg-laying cycle.

Tambja verconis (Polyceridae) and egg ribbon, Flinders Pier, Vic. Glenys Greenwood.

Nudibranchs and related molluscs are generally short-lived. Some small aeolids have a lifespan of 4–6 weeks. A few larger species may live 2–3 years, but most seem to be annual, appearing at much the same time each year.

Classification

Nudibranchs and related molluscs are a small part of the phylum Mollusca, class Gastropoda. They are closely related to garden snails and garden slugs, abalone, cowries, limpets, trumpet and helmet shells, and volutes, all of which have a muscular foot by which they crawl about.

The traditional classification of nudibranchs and related molluscs is very unstable. Major and minor changes continue to occur as new data from larger fields of study are published. The classification tabled here is more traditional, although some very recent name changes are recorded as comments against the relevant species. As discussed in ***What are nudibranchs and related molluscs?*** the reader can expect further significant revisions to the major groupings as DNA-based studies reveal more about the evolutionary relationships of nudibranchs and related molluscs.

More information on **systematics** and classification can be found among the references listed under ***Further Information***.

An outline of the classification of nudibranchs and related molluscs found in southern Australia

Order Acteonida

Bubble shells: shell external, tightly coiled and often strongly sculptured with spiral rows of pits in infaunal species; shell openly coiled and fragile in epifaunal species; animal with external non-retractile penis on right side of neck; operculum often present; hunts and eats polychaete worms.

Families represented in this book: Acteonidae, Aplustridae, Bullinidae and Ringiculidae.

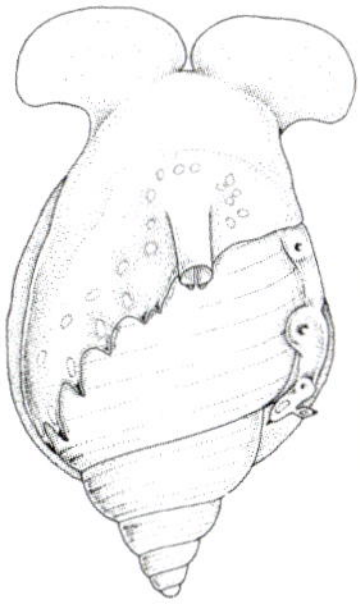

Ringicula sp. (Ringiculidae). Illustration: R. Plant from Beesley, Ross and Wells (1998).

Bullina lineata (Bullinidae). Leon Altoff.

Order Cephalaspidea

Bubble shells: shell external or reduced internal; animal with eversible penis within right side of head; operculum rarely present; gizzard plates often present; infaunal species usually cream all over, epifaunal species with strongly pigmented animal; active predators, some very selective.

Families represented in this book: Aglajidae, Bullidae, Cylichnidae, Diaphanidae, Gastropteridae, Haminoeidae, Ilbiidae, Philinidae, Retusidae, Runcinidae, Volvulellidae.

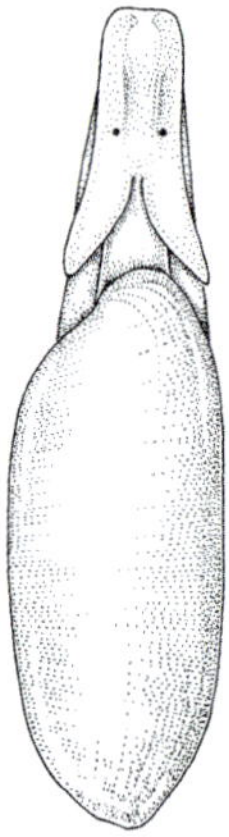

Volvulella rostrata (Volvulellidae), Oyster Harbour, WA. Illustration: R. Plant from Beesley, Ross and Wells (1998).

Diaphana tasmanica (Diaphanidae). Leon Altoff.

Order Rhodopemorpha

A little-known group of small to minute worm-like gastropods with seemingly few characteristics to separate the 10 or so known species, the majority of which occur in the Atlantic and Mediterranean. The animal is usually cylindrical or **fusiform**, and slightly flattened. The tail has a ventral adhesive gland. The head has slightly hollowed lateral cheeks with dorsal crests above. The head contracts into the body mass when disturbed. There are no jaws, radula or **gastral plates**.

There is only one family: Rhodopidae.

Not treated further in this book.

Rhodope sp. (Rhodopidae). Illustration: R. Plant from Beesley, Ross and Wells (1998).

Order Sacoglossa

Sap-sucking sea slugs: herbivores, mostly of green algae; shell when present thin and situated over **visceral mass**, animal often much larger than shell; animal with a pair of rhinophores on head; shell-less animals with a projecting, often wavy, fold (**parapodium**) along each side of body (Plakobranchidae), or with leaf-like or carrot-shaped cerata along each side of body.

Families represented in this book: Caliphyllidae, Hermaeidae, Juliidae, Limapontiidae, Oxynoidae, Plakobranchidae, Volvatellidae.

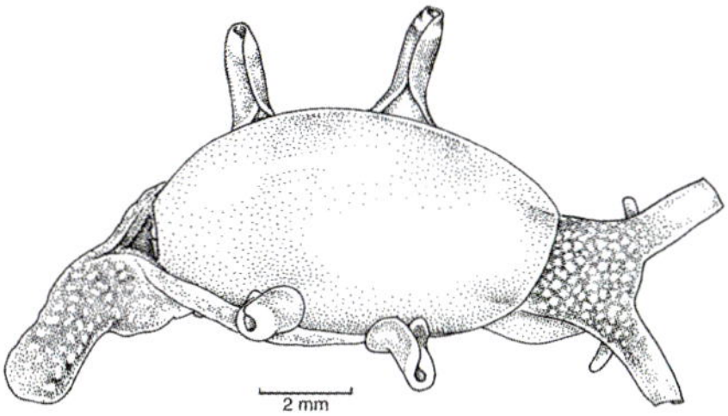

Roburnella wilsoni (Oxynoidae). Illustration: R. Plant from Beesley, Ross and Wells (1998).

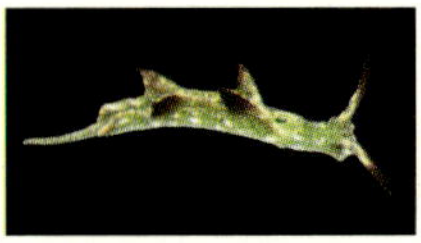

Elysia sp. RB2 (Plakobranchidae). Platon Vafiadis.

Ercolania boodleae (Limapontiidae). John Chuk.

Hermaea evelinemarcusae (Hermaeidae), Balnarring Beach, Vic. John Chuk.

Stiliger smaragdinus (Limapontiidae). Platon Vafiadis.

Order Anaspidea

Sea hares: shallow-water herbivores; shell when present usually fragile, internal, thin, towards rear of body over mantle cavity and gill; animal with oral tentacles and rhinophores on head, and **parapodia** each side of body, often with defensive purple gland associated with mantle cavity.

Families represented in this book: Akeridae and Aplysiidae.

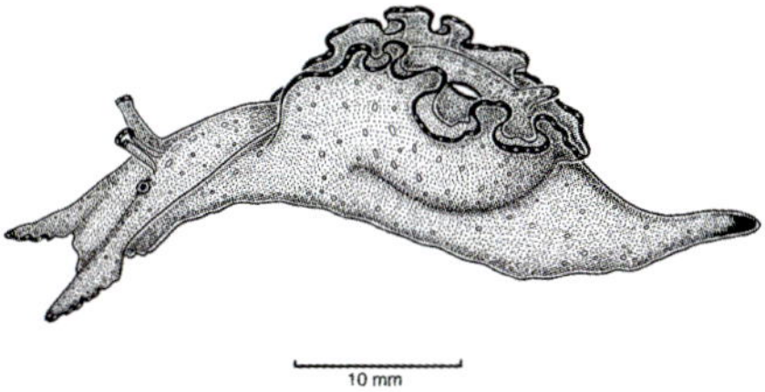

Aplysia parvula (Aplysiidae). Illustration: L. Newman from Beesley, Ross and Wells (1998).

Aplysia juliana (Aplysiidae). Leon Altoff.

Order Umbraculida

Umbrella shells: spongivores; very few species worldwide; shell external, flattened and limpet-like on top of larger animal, gill present on right side (*Tylodina*) or on right side and across front of animal below shell (*Umbraculum*).

Families represented in this book: Tylodinidae and Umbraculidae.

Tylodina corticalis (Tylodinidae), shell and animal. Illustration: R. Burn, from Burn (1960).

Tylodina corticalis (Tylodinidae). Platon Vafiadis.

Order Pteropoda

Sea butterflies: highly modified gastropods adapted to a planktonic existence in the open ocean, spending their entire lives suspended in the water column; foot developed into a pair of swimming wings.

Not treated further in this book.

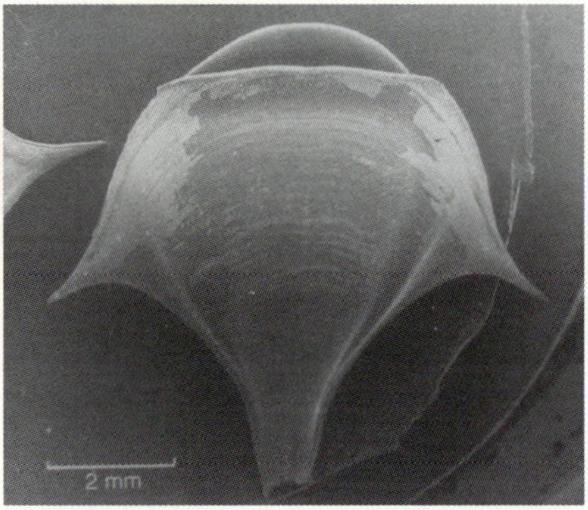

Creseis chierchiae (Cavolinidae). SEM Photo: L. Newman from Beesley, Ross and Wells (1998).

Order Acochlidia

Generally minute dwellers among sand grains, presently not known from southern Australia.

Not treated further in this book.

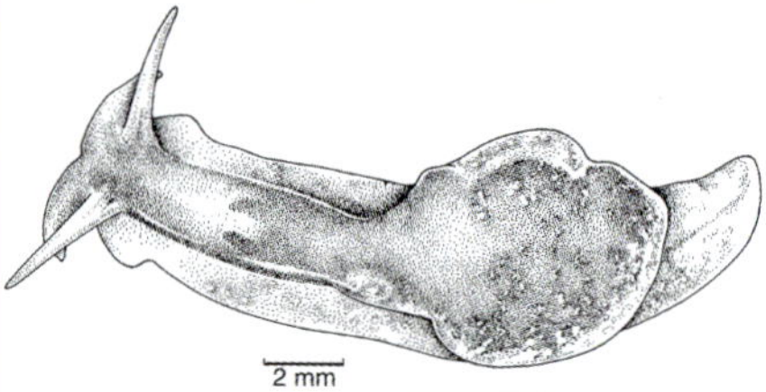

Acochlidium fijiense (Acochlidiidae). Illustration: R. Plant from Beesley, Ross and Wells (1998).

Order Pleurobranchida

Side-gilled slugs: spongivores, omnivores or ascidian eaters; shell when present internal, flattened, fragile, fingernail-shaped with small, coiled embryonic spire at one end; gill along right side of body; velum present across head, projecting forward of mantle; rhinophores projecting from below anterior mantle; larger species often strongly pustulose or tuberculose and brightly coloured.

One family represented in this book: Pleurobranchidae.

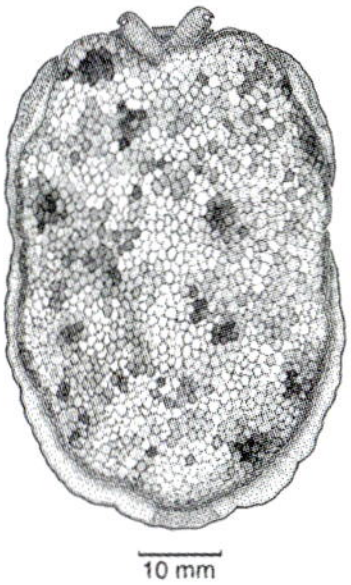

Pleurobranchus peronii (Pleurobranchidae). Illustration: L. Newman from Beesley, Ross and Wells (1998).

Pleurobranchaea maculata (Pleurobranchidae). Platon Vafiadis.

Order Nudibranchia

Nudibranchs: mantle with pair of rhinophores anteriorly, gills as a cluster posteriorly, laterally, or laterally below mantle overhang. The remaining groups treated below comprise the true nudibranchs.

Suborder Doridina

Dorids: generally spongivores but also prey on bryozoans, kamptozoans, ascidians; some are cannibalistic carnivores; mantle usually elliptical with overhang (**hyponotum**) all round, sometimes reduced to a small **velar** and **pallial** ridge, with or without projections; mantle surface smooth to highly ornamented with **pustules**, **papillae** or **tubercles**; gill cluster rear dorsal, rarely (tropical species) lateral along each side of body below mantle; oral tentacles present each side of mouth, usually hidden under mantle overhang.

Families represented in this book: Actinocyclidae, Aegiridae, Cadlinidae, Chromodorididae, Dendrodorididae, Discodorididae, Dorididae, Goniodorididae, Gymnodorididae, Onchidorididae, Polyceridae, Vayssiereidae.

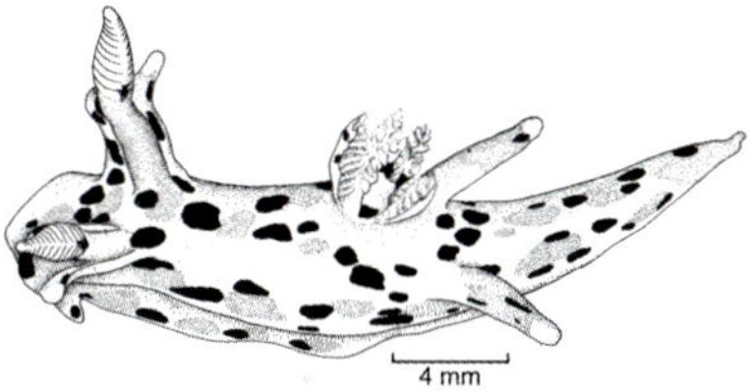

Thecacera pennigera (Polyceridae). Illustration: S. Weidland from Beesley, Ross and Wells (1998).

Okenia echinata (Goniodorididae). John Chuk.

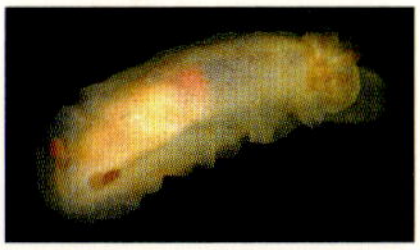

Dendrodoris maugeana (Dendrodorididae). Platon Vafiadis.

Crimora multidigitalis (Polyceridae). John Chuk.

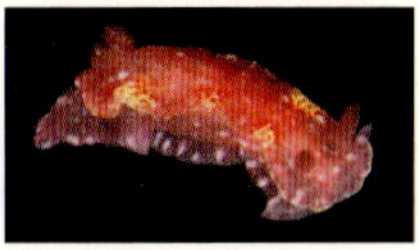

Thorunna arbuta (Chromodorididae). Platon Vafiadis.

Suborder Dendronotina
Dendronotids: eat hydroids, anemones, small crustaceans; elongate body; tree-like or carrot-shaped gill structures along each side of body formed from lateral fragmentation of mantle overhang; velum present across front of head; rhinophores projecting from high **sheaths**; anus opening right dorsally in front half of body (Dotidae, Tethydidae) or right laterally (other families).

Families represented in this book: Dendronotidae, Dotidae, Scyllaeidae, Tethydidae, Tritoniidae.

Doto ostenta (Dotidae). Illustration: S. Weidland from Beesley, Ross and Wells (1998).

Tritonia sp. RB3 (Tritoniidae). John Chuk.

Doto pita (Dotidae). John Chuk.

Scyllaea pelagica (Scyllaeidae). Mark Norman.

Melibe sp. RB1 (Tethydidae). Platon Vafiadis.

Suborder Arminina
Arminids: eat corals, soft corals, bryozoans; velum present across front of head; rhinophores without sheaths; anus opening right laterally, or posteriorly in the dorsal midline; mantle overhang with gills laterally below (Arminidae) or fragmented into tree-like or carrot-shaped projections of branchial capacity, the latter present across the head (Madrellidae, Proctonotidae).

Families represented in this book: Arminidae, Madrellidae, Proctonotidae.

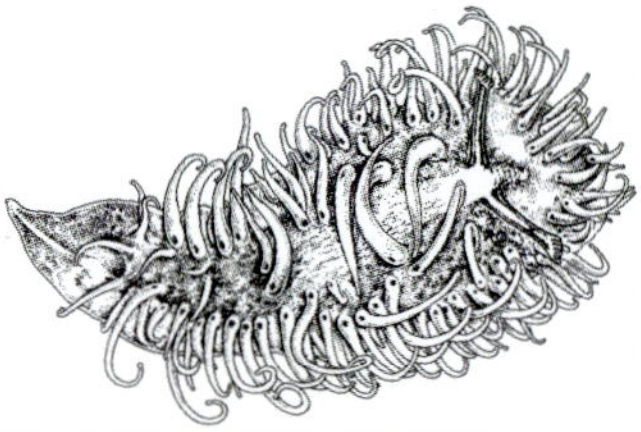

Madrella sanguinea (Madrellidae). Illustration: R. Plant from Beesley, Ross and Wells (1998).

Caldukia affinis (Proctonotidae). Leon Altoff.

Suborder Aeolidina

Aeolids: eat hydroids, anemones, soft coral, rarely bryozoans and the eggs of fish and nudibranchs; body slender, no mantle or overhang; oral tentacles on front of head; rhinophores without sheaths; anus opening right laterally or right dorsally; cerata along each side of body, tips of cerata with cnidosacs for storing nematocysts.

Families represented in this book: Aeolidiidae, Embletoniidae, Eubranchidae, Facelinidae, Fionidae, Flabellinidae, Glaucidae, Tergipedidae.

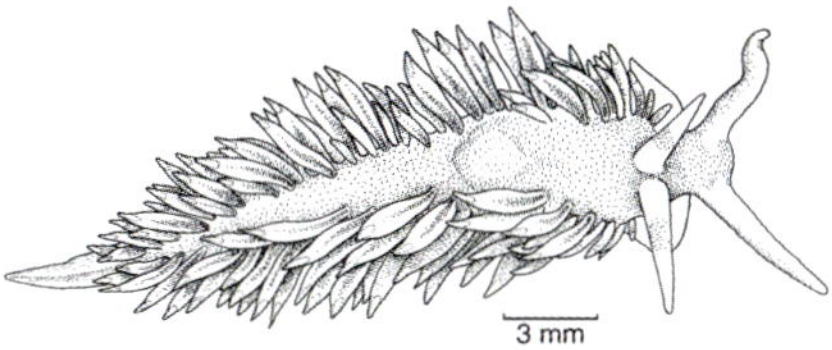

Anteaeolidiella sp. (Aeolidiidae). Illustration: S. Weidland from Beesley, Ross and Wells (1998).

Flabellina poenicia (Flabellinidae). John Chuk.

Glaucus atlanticus (Glaucidae). Bill Rudman, Australian Museum.

Cerberilla sp. RB1 (Aeolidiidae). John Chuk.

Austraeolis ornata (Facelinidae). Platon Vafiadis.

Where do they live?

Nudibranchs and related molluscs occur from the intertidal zone to very great depths (over 6000 m). Some can tolerate the reduced salinities of estuaries, and some very few tropical species have invaded the lower reaches of freshwater rivers and streams, but otherwise all species are fully marine.

Nudibranchs and related molluscs usually occur not far from their food. Intertidally, most seek protection from sunlight and exposure at low tide by hiding under rocks, in crevices or on algal holdfasts, or by burrowing into the sandy or muddy substrate. Early morning or late evening, even darkness, will find them crawling about in rock pools. In subtidal waters, nudibranchs and relatives are seen at all times, but more are active at night.

Shell-bearing species are for the large part infaunal, crawling through the top few millimetres of silty sand substrate. Larger species are more obvious as they plough along, forming a small hump at the end of their track. Infaunal species are usually creamy white in colour of both animal and shell. They can be very common but are rarely seen alive unless the substrate is fanned (which destroys visibility for some minutes) or run through a sieve to concentrate the specimens. Epifaunal species are usually pigmented in both animal and shell to match their exposed habitat.

The open ocean is also home to many species that are rarely seen by the shore observer or diver, except perhaps after many days of strong onshore winds and storms. The two suborders of sea butterflies (pteropods) occur in the water column offshore from south-eastern Australia: one with a shell (Thecosomata), the other without (Gymnosomata). The fragile shells of the former are common in deep-sea sediment samples, but the animals of both suborders need special shipboard sampling gear for retrieval. Two aeolid nudibranch genera live on the ocean surface and are sometimes washed ashore as the result of storms. The two species of *Glaucus*, kept afloat by ingested air bubbles, live upside down at the sea surface, using their cerata to row about hunting their food, which includes other floating animals such as Portuguese-man-of-war, by-the-wind-sailor and blue buttons. The one species of *Fiona* feeds upon goose barnacles that grow upon floating objects such as cuttle bones. Occasionally, live *Fiona* can be found among goose barnacles on newly washed-ashore flotsam.

What do they eat?

The vast majority of nudibranchs and related molluscs are carnivores and are to be found on or not far away from their food. The flattened, broadly oval dorids eat the tissue of sponges. Often their **notum** matches the sponge in colour and texture, making them difficult to distinguish from their food. The generally slender aeolids eat hydroids and anemones and are often difficult to see because their colours often blend into the mosaic of their food. Hydroids and anemones contain aggressive and defensive nematocysts in their tentacles. Aeolids are able to ingest nematocysts without triggering them, pass them through their digestive system, and store them for the aeolid's own defence in flask-like structures called cnidosacs in the tips of their cerata. Other forms are active predators of minute crustaceans, bryozoans, soft corals, nodding heads (kamptozoans), ascidians, nudibranchs, and even the eggs of nudibranchs and fish. A few, mainly of the genus *Pleurobranchaea*, are omnivorous, sweeping up all soft-bodied animals before them, including young of their own kind.

The herbivorous minority includes a number of the bubble shells, the sea hares, and the sap-sucking sacoglossans. Sea hares eat soft green and red seaweeds, often taking on body colouration to match. Some smaller species browse the surface of seagrasses, and others favour blue-green algae (also called **cyanobacteria**, which are toxic to humans). The digestive system of sacoglossans is equipped with blade-like radular teeth to pierce and slit the surface of green or red algae, and a muscular buccal pump by which the algal juices can be sucked out. Most sacoglossans have a green or red body colour matching that of their food. The diet of many sacoglossans is restricted to only one alga.

In addition, some sacoglossans sequester algal chloroplasts and 'farm' them within the digestive gland ductwork in their body and cerata. Chloroplasts are the subcellular structures that capture the sun's energy by photosynthesis, and they continue to provide an in-house energy source when consumed by the sacoglossan. A number of nudibranchs have the same capability, capturing protozoans called zooxanthellae, obtained from corals and other marine organisms that they have eaten.

What eats nudibranchs and related molluscs?

A shell into which an animal can retract for protection becomes largely unnecessary when the animal itself is armed with various forms of defensive glands. The shell-less nudibranchs have the greatest development of defensive glands. In dorids, these are either around the **notal margin**, or parts thereof, or spread over the whole notum. In aeolids, as well as glands in the cerata there is a cnidosac (storage flask) full of nematocysts in the tip of each ceras. These defences make nudibranchs highly unpalatable to most would-be predators.

However, some sea slugs actively seek out and eat other related species. Species of the cephalaspidean genus *Philinopsis* have no jaws, radula or gastral plates in the alimentary canal; their preferred food is other smaller bubble shells, including young of their own kind, which are sucked into the muscular pharynx and passed into the stomach. Here enzymes break down and digest the ingested animals, and the newly emptied shells are discharged via the intestine as part of the **faecal waste**.

Species of the doridoidean nudibranch genus *Gymnodoris* hunt and eat animals of the sacoglossan genus *Elysia*. *Gymnodoris alba*, a wide-ranging tropical and warm-temperate species occurring in the Bass Strait area, includes the aeolid nudibranch *Favorinus* sp. RB1 in its diet. *Favorinus* sp. RB1 in its turn devours the eggs of many other nudibranchs. This sequence of events is remarkably illustrated in the accompanying photograph, where it can be seen that some of the first laid eggs of *Doris cameroni* have hatched and released minute crawling young. Colour differences in the aeolids give some indication of the eggs being eaten: yellow from the yellow eggs of *Doris cameroni*, reddish in a newly arrived specimen that has previously been eating the reddish eggs of *Aplysia*, *Discodoris* or *Rostanga*.

Gymnodoris alba (Gymnodorididae) about to attack and eat *Favorinus* sp. RB1 (Facelinidae), which in turn is eating eggs of *Doris cameroni* (Dorididae), Blairgowrie Marina, Vic. John Chuk.

Favorinus sp. (Facelinidae) eating *Aplysia* eggs, Portsea Pier, Vic. John Chuk.

Nudibranch specimens with damaged edges to the notum or foot, or missing oral tentacles, rhinophores or cerata are often the result of encounters with the pincers (chelae) of crabs, snapping prawns or large shrimps, or from the tentative mouthing of smallish inquisitive fish.

Partnerships and introduced species

As a rule, nudibranchs and related molluscs do not share any commensal or symbiotic partnerships with other marine organisms, but they are extensively involved in mimicry of other organisms. They are all very dependent upon their food sources and, in fact, may never move away from them. Several south-eastern Australian species are considered to be introduced, but by what means remains a puzzle.

The nudibranch *Polycera hedgpethi* has a wide (nearly global) overseas distribution, having been described from California in 1964. It is a common, ever-present species in the **fouling community** on wharf piles in Port Phillip Bay, where it feeds upon the erect bryozoan *Bugula dentata*. The sacoglossan sea slug *Aplysiopsis formosa* is a North-east Atlantic and Caribbean species that feeds upon the green alga *Cladophora prolifera*. This alga was first noticed in Australia in the early 1950s and the sacoglossan at Port Phillip Heads in 1993; both remain common. Another sacoglossan, *Ercolania boodleae*, is a northern Pacific species that suddenly appeared in the Port Phillip Heads area in 2002, also feeding upon *Cladophora prolifera*. It too remains common.

About this guide

As noted throughout the introduction to this guide, the nudibranchs of southern Australia are a diverse group, with many new species still to be formally described. Most of the commonly seen species are included here; the author's aim is to extend this range—particularly via online resources—as images and samples are collected in the future.

A brief pictorial key to the higher groupings (orders and suborders) of southern Australian nudibranchs and related molluscs was given on pages 21-27. The following pages provide a brief description of the relevant orders and families, along with a summary of the key characteristics unique to individual species. Species that are yet to be documented in the **taxonomic** literature appear as a genus name followed by a species number, for example *Philine* sp. RB1. The numbering of species within a genus corresponds to the numbers allocated by the author in earlier reports. This guide also includes a number of species (both named or allocated a number) in excess of those listed in Burn (2006)—with the publication of the present guide, the author is now turning his efforts to more fully documenting these additional newly discovered species for a future revision of the Burn (2006) checklist.

Terms defined in the ***Glossary*** are printed in boldface type on first usage. Other words describing the animals in this field guide that may be less familiar are usually based on the names of the scientific groups to which they belong, e.g. 'dotid' indicates the organism belongs in the family Dotidae. If you refer back to the description of that group, you will know a lot about the organism.

Checklist of included species

Order Acteonida

Family Acteonidae

Pupa affinis (A. Adams, 1855)

Family Aplustridae

Hydatina physis (Linnaeus, 1758)

Family Bullinidae

Bullina lineata (Gray, 1825)

Family Ringiculidae

Ringicula doliaris Gould, 1860

Order Cephalaspidea

Family Aglajidae

Melanochlamys handrecki Burn, 2010
Melanochlamys queritor (Burn, 1957)
Noalda exigua (Hedley, 1912)
Philinopsis lineolata (H. Adams & A. Adams, 1854)
Philinopsis speciosa Pease, 1860
Philinopsis taronga (Allan, 1933)

Family Bullidae

Bulla quoyii Gray in Dieffenbach, 1843

Family Cylichnidae

Adamnestia arachis (Quoy & Gaimard, 1833)
Cylichna thetidis Hedley, 1903
Tornatina apicina Gould, 1859
Tornatina sp. RB1
Tornatina sp. RB2

Family Diaphanidae

Colpodaspis sp. RB1
Colpodaspis sp. RB2
Colpodaspis sp. RB3
Diaphana tasmanica (Beddome, 1883)

Family Gastropteridae

Sagaminopteron ornatum Tokioka & Baba, 1964
Siphopteron sp. RB1
Siphopteron sp. RB2
Siphopteron sp. RB3

Family Haminoeidae

Austrocylichna exigua (A. Adams, 1850)
Cylichnatys campanula Burn, 1978
Haminoea maugeansis Burn, 1966
Liloa brevis (Quoy & Gaimard, 1833)
Nipponatys tumidus Burn, 1978

Family Ilbiidae

Ilbia ilbi Burn, 1963

Family Philinidae

Philine angasi (Crosse in Crosse & Fischer, 1865)
Philine sp. RB1
Philine sp. RB2

Family Retusidae

Retusa atkinsoni (Tenison Woods, 1876)
Retusa pelyx Burn in Burn & Bell, 1974

Family Runcinidae

Runcina australis Burn, 1963
Runcina sp. RB1

Family Volvulellidae

Volvulella rostrata (A. Adams, 1850)

Order Sacoglossa

Family Caliphyllidae

Polybranchia pallens (Burn, 1957)

Family Hermaeidae

Aplysiopsis formosa Pruvot-Fol, 1953

Hermaea evelinemarcusae Jensen, 1993

Hermaea sp. RB1

Family Juliidae

Edenttellina typica Gatliff & Gabriel, 1911

Midorigai australis Burn, 1960

Tamanovalva babai Burn, 1965

Family Limapontiidae

Costasiella sp. RB1

Costasiella sp. RB2

Ercolania boodleae (Baba, 1938)

Ercolania margaritae Burn, 1974

Ercolania sp. RB2

Placida dendritica (Alder & Hancock, 1843)

Stiliger smaragdinus Baba, 1949

Family Oxynoidae

Oxynoe viridis (Pease, 1861)

Roburnella wilsoni (Tate, 1889)

Family Plakobranchidae

Elysia coodgeensis Angas, 1864

Elysia furvacauda Burn, 1958

Elysia maoria Powell, 1937

Elysia sp. RB1

Elysia sp. RB2

Elysia sp. RB3

Family Volvatellidae

Ascobulla fischeri (A. Adams & Angas, 1864)

Order Anaspidea

Family Akeridae

Akera soluta (Gmelin, 1791)

Family Aplysiidae

Aplysia dactylomela Rang, 1828

Aplysia juliana Quoy & Gaimard, 1832

Aplysia parvula Mörch, 1863

Aplysia sowerbyi Pilsbry, 1895

Aplysia sydneyensis Sowerby I, 1869

Bursatella sp. RB1

Dolabrifera brazieri Sowerby II, 1870

Order Umbraculida

Family Tylodinidae

Tylodina corticalis (Tate, 1889)

Family Umbraculidae

Umbraculum umbraculum (Lightfoot, 1786)

Order Pleurobranchida

Family Pleurobranchidae

Berthellina citrina (Rüppell & Leuckart, 1828)

Berthella medietas Burn, 1962

Berthella serenitas Burn, 1962

Pleurobranchaea maculata (Quoy & Gaimard, 1832)

Pleurobranchus hilli (Hedley, 1894)

Order Nudibranchia

Suborder Doridina

Family Actinocyclidae

Hallaxa michaeli Gosliner & Johnson, 1994

Family Aegiridae

Aegires exeches Fahey & Gosliner, 2004

Family Cadlinidae

Cadlina tasmanica Rudman, 1990

Family Chromodorididae

Ceratosoma amoenum (Cheeseman, 1886)

Ceratosoma brevicaudatum Abraham, 1876

Chromodoris alternata (Burn, 1957)

Chromodoris ambigua Rudman, 1987

Chromodoris tasmaniensis Bergh, 1905

Diversidoris sulphurea (Rudman, 1986)

Doriprismatica kulonba (Burn, 1966)

Goniobranchus epicurius (Basedow & Hedley, 1905)

Goniobranchus tinctorius (Rüppell & Leuckart, 1828)

Hypselodoris bennetti (Angas, 1864)

Mexichromis macropus Rudman, 1983

Noumea aureopunctata Rudman, 1987

Noumea closeorum Rudman, 1986

Noumea haliclona (Burn, 1957)

Noumea verconis (Basedow & Hedley, 1905)

Thorunna arbuta (Burn, 1961)

Thorunna perplexa (Burn, 1957)

Family Dendrodorididae

Dendrodoris albopurpura Burn, 1957

Dendrodoris arborescens (Collingwood, 1881)

Dendrodoris aurea (Quoy & Gaimard, 1832)

Dendrodoris maugeana Burn, 1962

Dendrodoris nigra (Stimpson, 1855)

Doriopsilla carneola (Angas, 1864)

Doriopsilla peculiaris (Abraham, 1877)

Family Discodorididae

Alloiodoris marmorata Bergh, 1904

Discodoris paroa Burn, 1969

Hoplodoris nodulosa (Angas, 1864)

Jorunna hartleyi (Burn, 1958)

Jorunna pantherina (Angas, 1864)

Jorunna ramicola Miller, 1996

Jorunna sp. RB1

Paradoris dubia (Bergh, 1904)

Platydoris galbana Burn, 1958

Rostanga bassia Rudman & Avern, 1989

Rostanga bifurcata Rudman & Avern, 1989

Rostanga calumus Rudman & Avern, 1989

Rostanga crawfordi (Burn, 1969)

Sclerodoris tarka Burn, 1969

Sclerodoris trenberthi (Burn, 1962)

Sclerodoris sp. RB1

Tayuva lilacina (Gould, 1852)

Thordisa sanguinea Baba, 1955

Thordisa verrucosa (Crosse in Angas, 1864)

Trippa albata Burn, 1962

Family Dorididae

Aphelodoris berghi Odhner, 1924
Aphelodoris rossquicki Burn, 1966
Aphelodoris varia (Abraham, 1877)
Aphelodoris sp. RB1
Doris cameroni (Allan, 1947)
Doris chrysoderma Angas, 1864
Doris flabellifera Cheeseman, 1881
Doris wellingtonensis Abraham, 1877

Family Goniodorididae

Ancula mapae (Burn, 1961)
Ancula sp. RB1
Ancula sp. RB2
Goniodoridella savignyi Pruvot-Fol, 1933
Goniodoris meracula Burn, 1958
Okenia echinata Baba, 1949
Okenia mija Burn, 1967
Okenia plana Baba, 1960
Okenia rhinorma Rudman, 2007
Okenia sp. RB1
Okenia sp. RB2
Okenia sp. RB3
Trapania aureopunctata Rudman, 1987
Trapania benni Rudman, 1987
Trapania brunnea Rudman, 1987

Family Gymnodorididae

Gymnodoris alba (Bergh, 1877)
Gymnodoris arnoldi (Burn, 1957)
Paliolla cooki (Angas, 1864)

Family Onchidorididae

Acanthodoris metulifera Bergh, 1905
Acanthodoris nanega Burn, 1969
Corambe sp. RB1
Diaphorodoris sp. RB1
Onchidoris maugeansis (Burn, 1958)

Family Polyceridae

Crimora multidigitalis (Burn, 1957)
Kaloplocamus ramosus (Cantraine, 1835)
Polycera capensis Quoy & Gaimard, 1824
Polycera hedgpethi Er. Marcus, 1964
Polycera janjukia Burn, 1962
Polycera melanosticta Miller, 1996
Polycera parvula (Burn, 1958)
Polycera sp. RB1
Polycera sp. RB2
Tambja verconis (Basedow & Hedley, 1905)
Tambja sp. RB1
Tambja sp. RB2
Thecacera pennigera (Montagu, 1815)

Family Vayssiereidae

Vayssierea caledonica Risbec, 1928

Suborder Dendronotina

Family Dendronotidae

Dendronotus sp. RB1
Dendronotus sp. RB2

Family Dotidae

Doto ostenta Burn, 1958
Doto pita Er. Marcus, 1955
Doto sp. RB1
Doto sp. RB2
Doto sp. RB3
Doto sp. RB4

Family Scyllaeidae

Crosslandia viridis Eliot, 1902
Scyllaea pelagica Linnaeus, 1758

Family Tethydidae

Melibe australis Angas, 1864
Melibe maugeana Burn, 1960
Melibe sp. RB1
Melibe sp. RB2
Melibe sp. RB3

Family Tritoniidae

Marioniopsis platyctenea Willan, 1988
Marianina rosea (Pruvot-Fol, 1930)
Tritonia sp. RB1
Tritonia sp. RB2
Tritonia sp. RB3
Tritonia sp. RB4

Suborder Arminina

Family Arminidae

Armina sp. RB1
Dermatobranchus rubidus (Gould, 1852)

Family Madrellidae

Madrella sanguinea (Angas, 1864)
Madrella sp. RB1

Family Proctonotidae

Caldukia affinis (Burn, 1958)
Janolus hyalinus (Alder & Hancock, 1854)
Janolus sp. RB1
Janolus sp. RB2
Janolus sp. RB3
Janolus sp. RB5

Suborder Aeolidina

Family Aeolidiidae

Aeolidiella drusilla Bergh, 1900
Anteaeolidiella lurana (Marcus & Marcus, 1967)
Baeolidia australis (Rudman, 1982)
Baeolidia macleayi (Angas, 1864)
Burnaia helicochorda (Miller, 1988)
Cerberilla incola Burn, 1974
Cerberilla sp. RB1

Family Embletoniidae

Embletonia gracilis Risbec, 1928

Family Eubranchidae

Eubranchus rubeolus Burn, 1964
Eubranchus sp. RB1
Eubranchus sp. RB2
Eubranchus sp. RB3
Eubranchus sp. RB6
Eubranchus sp. RB9

Family Facelinidae

Austraeolis ornata (Angas, 1864)
Cratena lineata (Eliot, 1905)
Echinopsole breviceratae Burn, 1962
Facelina hartleyi Burn, 1962
Facelina newcombi (Angas, 1864)
Facelina sp. RB1
Facelina sp. RB2
Facelina sp. RB3
Favorinus sp. RB1
Palisa sp. RB1
Phyllodesmium macphersonae (Burn, 1962)
Phyllodesmium serratum (Baba, 1949)

Family Fionidae

Fiona pinnata (Eschscholtz, 1831)

Family Flabellinidae

Flabellina poenicia (Burn, 1957)

Flabellina sp. RB1

Flabellina sp. RB2

Flabellina sp. RB4

Tularia bractea (Burn, 1962)

Family Glaucidae

Glaucus atlanticus Forster, 1777

Glaucus marginatus (Reinhardt in Bergh, 1864)

Family Tergipedidae

Tergipes sp. RB1

Trinchesia anulata (Baba, 1949)

Trinchesia catachroma (Burn, 1963)

Trinchesia kuiterorum Rudman, 1981

Trinchesia ornata (Baba, 1937)

Trinchesia sororum Burn, 1964

Trinchesia thelmae (Burn, 1964)

Trinchesia viridiana (Burn, 1962)

Trinchesia sp. RB1

Trinchesia sp. RB2

Trinchesia sp. RB3

Trinchesia sp. RB5

Trinchesia sp. RB7

Trinchesia sp. RB10

Trinchesia sp. RB11

Trinchesia sp. RB14

Nudibranchs and related molluscs

Order Acteonida

The families of this order are united mainly by characteristics of the animal, for example the presence of an external pallial penis situated just within the mantle cavity. Shells may be thick or thin, with a relatively small or very large aperture, and an operculum is present in two of the three families.

Pupa affinis

Pupa affinis is widespread in the tropical and warm-temperate Indo-Pacific and is a common subtidal inhabitant of the estuaries of southern New South Wales. At times it may occur in similar environments in far eastern Victoria, and it is also known from shallow-water dredgings from off the coast of Lakes Entrance.

Family Aplustridae

A family of usually flamboyant, brightly pigmented, epifaunal species, common in tropical and warm-temperate waters, with one species rarely appearing in far eastern Victoria. The shell has no spire, is light and thin, and the aperture is large. The large animal has a well-developed headshield, the anterior corners extended into deeply folded, hollow, tentacle-like structures. There is no operculum.

Family Acteonidae

Acteonids have thick shells sculptured with close spirals of **punctae**, and the spire is short. The animal can retract within the shell and close the aperture with its horny operculum. There are no intertidal or shallow subtidal species in Victorian waters. Those that are recorded for Victoria are all from deeper water, where they burrow within the top few millimetres of bottom substrate.

Pupa affinis (Acteonidae), Beachmere, Moreton Bay, Qld. Leon Altoff.

Hydatina physis

The shell is white in colour with (usually numerous) brown bands. The animal is pink or dusky pink, with white or bluish-white margins to the foot, head-shield and posterior mantle folds. A pair of black eyes is visible on the head.

Size of shell: 30 mm.

Habitat and range: 0–10 m depth; epifaunal on sandy mud near seagrass beds. Rarely in the intertidal zones of far eastern Vic. Otherwise circumglobal in tropical and temperate coastal waters.

Family Bullinidae

The shell of bullinid species has a short spire as in acteonids, but is light and thin as in aplustrids. It is usually white in colour with a red spiral and axial bands. The animal is usually translucent white with brightly pigmented margins. An operculum is present, only just filling the shell aperture. Bullinids are epifaunal animals, often seen crawling in the lower intertidal pools in New South Wales.

Bullina lineata

Shell white with well-spaced axial and spiral red lines, sculptured with spiral rows of punctae. Animal translucent whitish; margins of foot, headshield and posterior mantle folds a brilliant electric blue. A pair of black eyes is visible on the head.

Size of shell: 10–12 mm.

Habitat and range: 0–45 m depth; epifaunal on sandy mud, seagrass and algae in rock pools and estuaries. Common intertidally in southern NSW, uncommon in eastern Bass Strait; otherwise, tropical and temperate Indo-Pacific.

Hydatina physis (Aplustridae), Bastion Point, Mallacoota, Vic. Leon Altoff.

Bullina lineata (Bullinidae), Bastion Point, Mallacoota, Vic. Leon Altoff.

Family Ringiculidae

Ringiculids have a stout shell with a short spire, a large aperture with a thickened outer lip, and very strong columellar teeth. The shell is usually creamy or greyish white. The animal appears to have no lateral division between foot and body, and the rear end of the head is formed into a high, cylindrical, exhalent siphon. Ringiculids are subtidal to deep-water infaunal species. One species of *Ringicula* lives in fine sediment around *Zostera* and *Halophila* beds growing in the shallow, subtidal region of the estuaries of southern New South Wales, and it may at times occur in similar habitats in far eastern Victoria. Victorian records are from deep water and need to be confirmed.

Ringicula doliaris (Ringiculidae), Grotto Point, Sydney Harbour, NSW. Bill Rudman, Australian Museum.

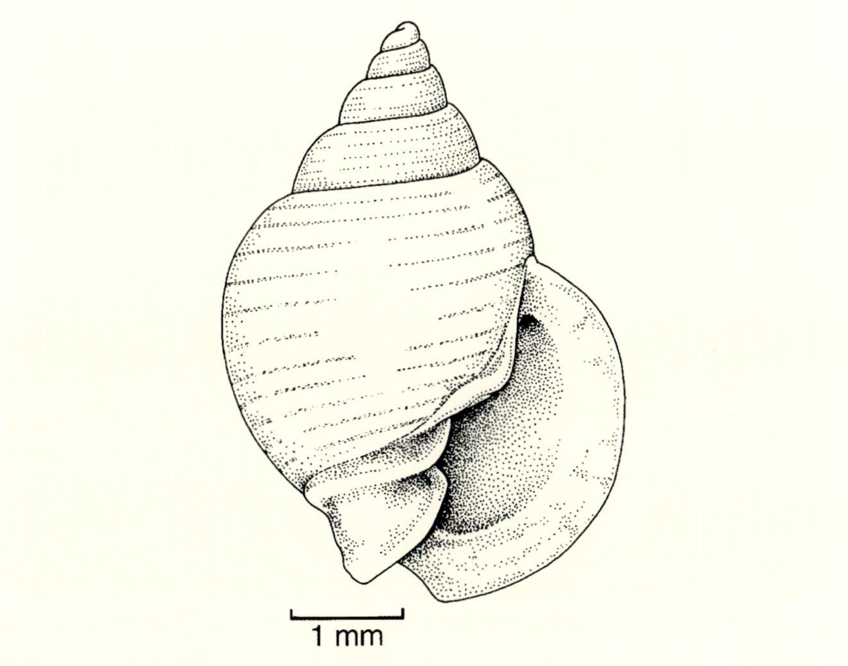

Ringicula sp. (Ringiculidae). Illustration: R. Plant from Beesley, Ross & Wells (1998).

Order Cephalaspidea

All cephalaspideans have a shell. Infaunal species usually have a cylindrical external shell into which the small animal can wholly retract. The animals of these species are uniformly whitish cream. Epifaunal species usually have a reduced, thin, fragile internal shell offering seemingly little protection to the visceral mass over which it lies. Epifaunal animals are often brightly pigmented or patterned, but exceptions abound. Most species have a radula and gastral plates, but some have one or the other, or even neither. Many cephalaspideans are extremely common in intertidal and shallow subtidal waters.

Melanochlamys handrecki

A small, flattened species, light to dark grey in colour on dorsal surfaces, greyish white below; parapodia not reaching high or tightly up the body sides; oral lobes with hyaline sensory papillae that project when the animal is searching for its polychaete food.

Size: 15–30 mm.

Habitat and range: lowest intertidal to 3 m depth; crawling through fine silt in areas rich in small polychaetes. Vic. and Tas.

Melanochlamys queritor

A small, cylindrical species, usually black in colour, though it varies from blotchy cream and brown to reddish brown. The parapodia are held close to the body, and there is a small notch between the short posterior tail folds. Readily separated from *Melanochlamys handrecki* by the cylindrical body, with parapodia held close to the body.

Size of animal: 10–20 mm, exceptionally 34 mm.

Habitat and range: intertidal to 6 m depth; usually crawling among seagrass and algae seeking its polychaete prey. NSW to southern WA, including Tas.

Family Aglajidae

Members of this family lack both radula and gastral plates, though there are one or two exceptions. Aglajids are elongate, cylindrical animals, rather soft and flabby, with a thin parapodial fold each side of the body. A shell, in some species resembling only the top quarter of a *Bulla* shell, is enclosed within the hind part of the visceral mass, though rarely it may be partly exposed. Some species have pads of **extensile setae** each side of the mouth that are used to search out food. Many species have a pair of fleshy trailing projections behind the animal, the left one being longer.

Aglajids are predatory carnivores, actively seeking out their food animals, which they ingest whole by means of a strongly muscular pharynx.

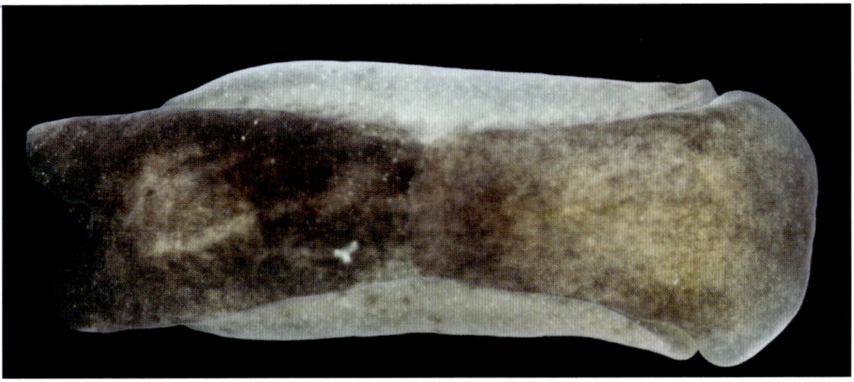

Melanochlamys handrecki (Aglajidae), Shallow Inlet, Vic. Leon Altoff.

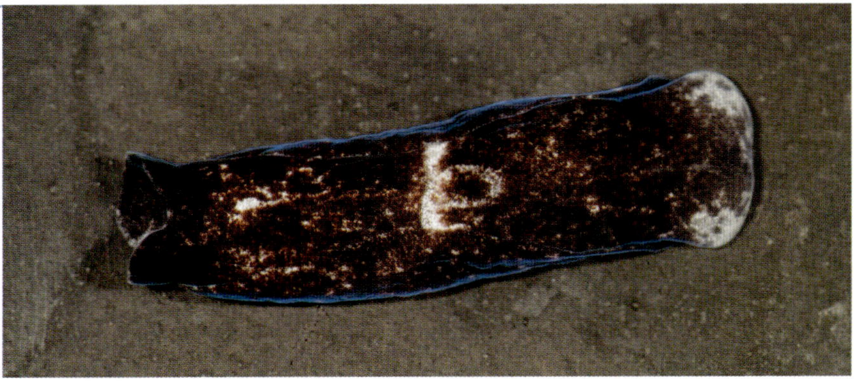

Melanochlamys queritor (Aglajidae), Flinders, Mushroom Reef, Vic. Leon Altoff.

Noalda exigua

One of the few species in the family Aglajidae with a partly exposed shell. Animal very small, white with yellow blotches along margins, left tail longer. Shell white, globose, with two red spiral bands, both visible in the exposure.

Size of animal: 5 mm, usually 3–4 mm.

Habitat and range: intertidal to 37 m depth; intertidally found among the wire weed *Amphibolis antarctica*. Central NSW, Vic., Tas. and SA.

Philinopsis lineolata

Immediately separated from the other two local species of *Philinopsis* by the narrow stripes of dark yellow transversing the whole black or bluish-black body. The head of this species is much narrower and higher than its local congeners.

Size of animal: 25–45 mm in Vic., to 70 mm in warmer waters.

Habitat and range: intertidal to 60 m depth; on sandy-mud bottoms where it hunts for polychaete worms. All Australian states (except Tas.); NZ and tropical and temperate Indo-Pacific.

Philinopsis speciosa

A warmer-water visitor; occasionally seen in the embayments of Victoria's central coastline. An elongate cylindrical animal with the posterior end of the headshield abruptly raised (like a broad shark fin). The parapodia are thin and often held a little away from the body. Usually bluish black in colour with red and/or yellow submarginal stripes on the headshield, tail folds and parapodia.

Size of animal: 20–34 mm in Vic. water, in excess of 100 mm elsewhere.

Habitat and range: intertidal to 20 m depth; usually on soft, muddy-sand sediments, where it hunts for bubble shells. Indo-Pacific tropical and temperate waters, rarely seen in Vic. Previously known as *Philinopsis cyanea*.

Noalda exigua (Aglajidae), Anderson Point, Portland, Vic. Leon Altoff.

Philinopsis lineolata (Aglajidae), Governor Reef, Indented Heads, Port Phillip Bay, Vic. Trevor McMurrich.

Philinopsis speciosa (Aglajidae), Portarlington, Port Phillip Bay, Vic. Trevor McMurrich.

Philinopsis taronga

In form, although not in colour, *Philinopsis taronga* is very similar to *P. speciosa*, but it is much more common. The posterior end of the headshield is usually flat or raised very little. The colour is lighter or darker brown with creamy-yellow or golden submarginal stripes on the head and parapodia, often with cream speckling all over.

Size of animal: 20–35 mm in Vic., up to 80 mm elsewhere.

Habitat and range: intertidal to 150 m depth; usually on soft, muddy-sand sediments, hunting for small bubble shells. NSW to Vic. and Tas.; also NZ.

Family Bullidae

This family includes the largest of the bubble shells. The shell is broadly to narrowly ovoid in shape; the spire is depressed, leaving only an apical umbilicus; the aperture is large, and the shell smooth and often glossy. The animal can be contained within the shell; the pharynx has a radula, and the gizzard both gastral and accessory plates. Bullids are herbivores and are found mostly in soft sediments near seagrass beds, from whence they emerge at night to forage.

Bulla quoyii

The ovoid shell is usually dull brownish, sometimes with bluish or greyish patches or bands. The animal is yellowish orange in colour, often with small, creamy speckles. The anterior corners of the head are formed into short, cylindrical tentacles; the parapodia are small, protecting only the neck and front of the shell.

Size of shell: to 45 mm.

Habitat and range: intertidal to 20 m depth; usually near seagrass beds on soft-bottom shores. All Australian states except the tropical north; also North Island of NZ.

Philinopsis taronga (Aglajidae), Bellerive Yacht Club, Tas. Sarah Quine.

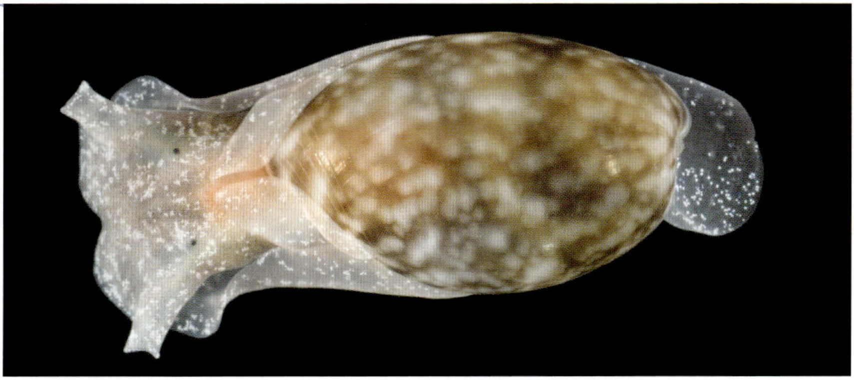

Bulla quoyii (Bullidae), Clifton Springs, Port Phillip Bay, Vic. Leon Altoff.

Family Cylichnidae

The shells in this family are either cylindrical or barrel-shaped, usually with a depressed or short spire, and the aperture is as long or almost as long as the shell. All species are infaunal, and this habitat is reflected in the drab-coloured shells and animals. The animals are small, about one-quarter of the shell length, with a pair of rounded lobes projecting rearwards on to the front of the shell. As a general rule, cylichnids have three gastral plates (usually a mirror image pair and one smaller to much smaller) for crushing and grinding food. The radula has at least a large sickle-shaped, denticulate lateral tooth each side of the radular row midline, and marginal teeth may also be present. Cylichnids are often common on intertidal and estuarine mudflats, and there are many subtidal and deeper-water species, both named and unnamed.

Some authors treat all species of *Tornatina* as belonging instead to the genus *Acteocina*, however consistent differences in radula and gastral plate morphology suggest that Indo-Pacific species represent a separate clade, here accepted as *Tornatina*.

Adamnestia arachis

The shell of this species is often invested with a brownish **periostracum**, hence the name *arachis* indicating its resemblance to the groundnut (or peanut). The shell is cylindrical, slightly narrower at midlength, the spire depressed and the anterior **columella** strongly twisted. The animal is creamy yellow.

Size of shell: 30–40 mm, usually smaller.

Habitat and range: 0–200 m depth; infaunal in mud and sandy mud. Southern Australia from Qld to WA.

Cylichna thetidis is a similar but smaller (10–15-mm) species with the same depth range and distribution. Its shell is more cylindrical, the posterior end of the aperture more open.

Adamnestia arachis (Cylichnidae), Port Welshpool, Vic. Platon Vafiadis.

Tornatina apicina

The small, cylindrical shell has a short, tapering spire, the **whorls** of which are separated each from the next by a deep, narrow furrow with a sharp outer edge. Shell white and glossy, in life often with a short green sprig of alga attached to the spire tip. Animal cream.

Size of shell: 4–5 mm.

Habitat and range: 0–150 m depth; in sandy mud, common in estuaries. NSW, Tas. and Vic.

There are at least five other species of *Tornatina* in Victorian waters. These can be separated by the shell surface sculpture, the size of the **protoconch** at the tip of the spire, and the shape of the furrow separating the spire whorls. There are also differences in the shape of the gastral plates and radula teeth.

Tornatina sp. RB1

Shell small and cylindrical with a rounded shoulder; protoconch larger than in *Tornatina apicina*. Shell semitranslucent white with opaque white patches. Animal cream.

Size of shell: 4–5 mm.

Habitat and range: 0–20 m depth; in sandy patches and among coralline algae on open coastal reefs. Vic. and SA.

Tornatina sp. RB2

Shell small, oval, wider posteriorly and spire very short; protoconch very small, often eroded. Shell generally yellowish, often with yellowish-orange periostracum. Animal yellowish cream.

Size of shell: 3–4 mm.

Habitat and range: 0–5 m depth; in muddy sand in shallow estuarine and well-protected bays. Vic. and SA.

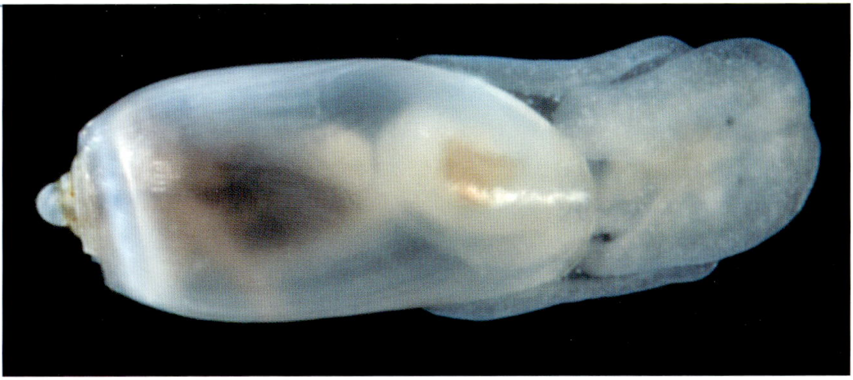

Tornatina apicina (Cylichnidae), Yeerung River, Vic. Leon Altoff.

Tornatina sp. RB1 (Cylichnidae), Point Addis, Vic. Leon Altoff.

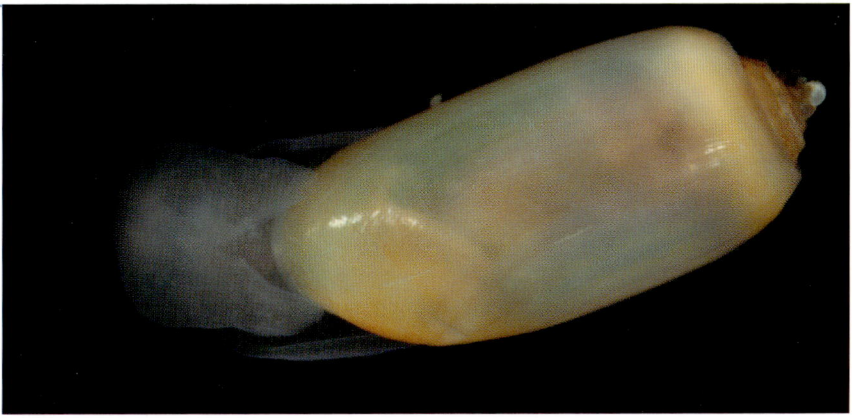

Tornatina sp. RB2 (Cylichnidae), Shallow Inlet, Vic. Leon Altoff.

Family Diaphanidae

Generally small animals with a thin, fragile, transparent shell into which the body can retract. The head is wider anteriorly, with a furrowed margin. The foot is short in those species with an external shell, sometimes indented in the rear margin. In species with an internal shell, the foot is long and slender and much wider in front. There are several undescribed species in Victorian waters.

Colpodaspis sp. RB2

Shell internal, white, broadly conical with a very short spire and a large aperture. Dark-brown or reddish-brown mantle covering the shell, a siphonal fold present on rear right side. Head slender with stout cylindrical tentacles in front, usually lighter brown than mantle. Foot broadly expanded in front and opaque white in colour; tail long and tapering (although partly retracted in the photo), varying from opaque white to dark grey.

Size of crawling animal: 6–7 mm, usually smaller.

Habitat and range: 0–95 m depth; intertidally associated with algae and the seagrass *Amphibolis*. Vic. and SA.

Two other undescribed species occur in Victorian waters: *Colpodaspis* sp. RB1 has a dark-blue mantle and no siphonal fold; *Colpodaspis* sp. RB3 has a pale-blue mantle and it does have a siphonal fold.

Diaphana tasmanica

Shell external, minute, **truncate** behind and slightly wider in front, reddish brown for rear half, more yellowish brown in front; animal white; black eyes prominent (although these are not visible in the photograph below).

Size of shell: 2 mm.

Habitat and range: intertidal and shallow subtidal; associated with filamentous and other brown algae. Vic. and Tas.

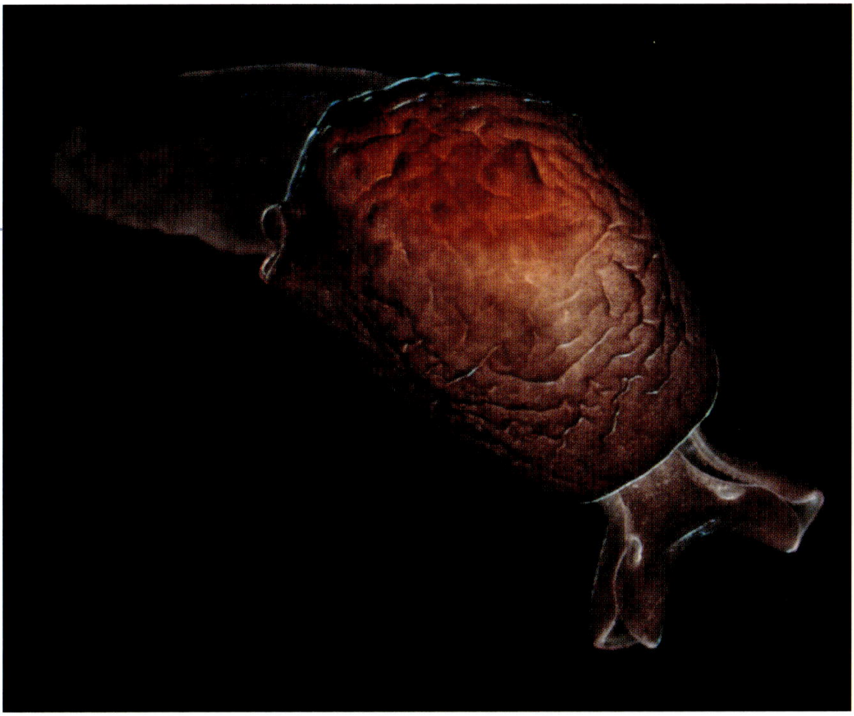

Colpodaspis sp. RB2 (Diaphanidae), Sisters Point, east of Port Fairy, Vic. Leon Altoff.

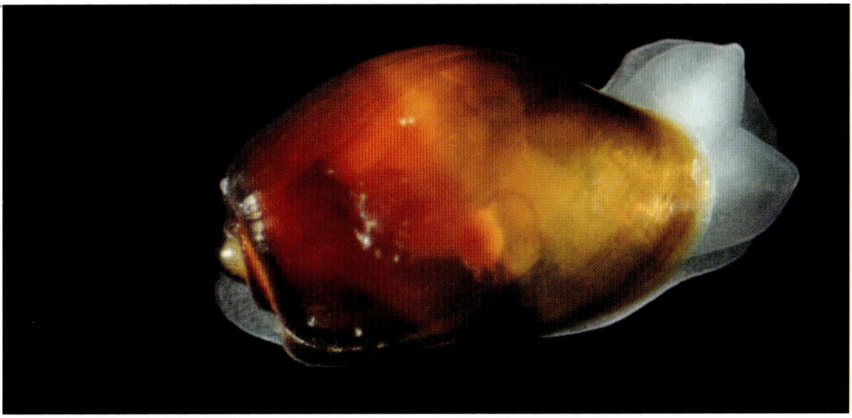

Diaphana tasmanica (Diaphanidae), Sisters Point, east of Port Fairy, Vic. Leon Altoff.

Family Gastropteridae

The very wide parapodia, the ability of many species to swim, the general lack of a shell except in the earliest stages of development, and the very reduced mantle cavity where the gill has moved forward and is exposed, separate gastropterids from the allied philinids and aglajids. Most species have the posterior end of the headshield narrowed, highly raised and curved forward, sometimes level with or even in front of the anterior edge of the head. The animal possesses a radula, but not gastral plates. Gastropterids are renowned for their striking colours and patterns, especially among the smaller species, of which local waters boast several undescribed species.

Sagaminopteron ornatum

The bat-wing slug is well known to local divers. The bright reddish-pink or purple body with yellow, orange or cream margins to the headshield, parapodia and the fleshy protuberances on the dorsal rear right of the visceral mass is most distinctive. A small, spoon-shaped internal shell has been reported for this species.

Size of animal: 10–25 mm.

Habitat and range: usually subtidal to 25 m depth; often associated with a creamy-grey sponge upon which it feeds. All Australian states; tropical and temperate Indo-Pacific.

Siphopteron sp. RB1

A small species with narrow parapodia; incapable of swimming. Orange in colour, sometimes with yellow patches on the head and parapodia. Head siphon and short visceral **flagellum** black tipped.

Size: 4 mm.

Habitat and range: 0–20 m depth; crawling on algae in low intertidal rock pools. Vic. and Tas.

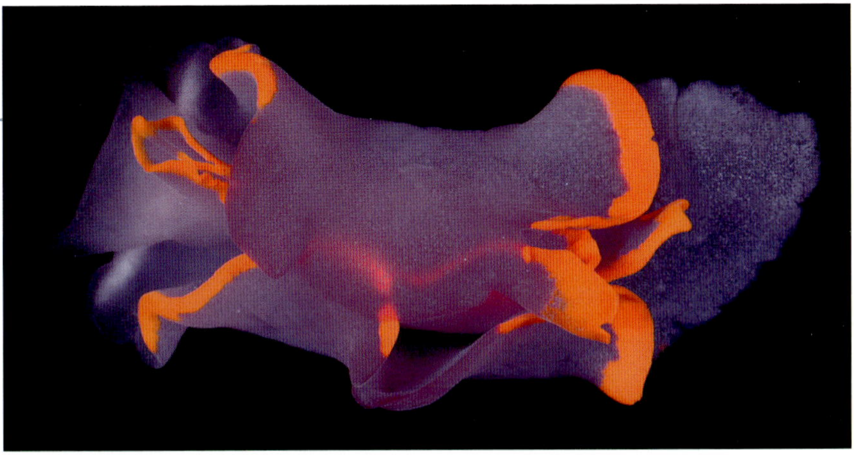

Sagaminopteron ornatum (Gastropteridae), Blairgowrie Marina, Vic. John Chuk.

Siphopteron sp. RB1 (Gastropteridae), West Head, Flinders, Vic. John Chuk.

Siphopteron sp. RB2

Slightly larger than *Siphopteron* sp. RB1, with wider parapodia used for brief bursts of swimming. Bright orange red in colour with irregular, more or less longitudinal, bluish lines on head, visceral mass, parapodia and foot. Head siphon and short visceral flagellum tipped bluish black.

Size: 5–6 mm.

Habitat and range: 0–20 m depth; crawling on algae in low intertidal rock pools, subtidally associated with the green alga *Cladophora prolifera*. Vic. and southern WA.

Siphopteron sp. RB3

A strikingly patterned animal; body overall brown with a tracery of bluish lines enclosing brown, polygonal patches, each with an orange central patch. Bluish-white patches are present at each anterior corner of the head and along the centreline of the tail. An irregular, transverse patch of white marks the top of the visceral mass and the base of the short, conical flagellum. Head siphon and visceral flagellum brown tipped.

Size: 4–6 mm.

Habitat and range: intertidal; associated with the seagrass *Halophila australis*. Vic.

Siphopteron sp. RB2 (Gastropteridae), Portarlington, Vic. Trevor McMurrich.

Siphopteron sp. RB3 (Gastropteridae), San Remo, Vic. Platon Vafiadis.

Family Haminoeidae

Another family of herbivorous species, the shell of which is ovoid or narrowed behind and into which the animal can retract. Haminoeids have both radula and gastral plates, and the latter generally have transverse crests of projecting elements. Most species are infaunal, burrowing through the fine top sediments of the substrate. Others are epifaunal, crawling on their algal foods. Animals that burrow are usually dull in colouration, whereas those that crawl epifaunally are often brightly pigmented. The inner whorls of the shell in all species of this family are constantly resorbed, leaving only about one-and-a-half whorls of shelly material to protect the animal. The parapodia are reduced to small, triangular folds, protecting only the neck of the animal and the anterior edge of the shell at the entry to the mantle cavity.

As a general rule, haminoeids live only in the intertidal or shallow subtidal zones. Many species are nocturnally active.

Austrocylichna exigua

This species is characterised by a slender, narrowly ovoid, cream-coloured shell with a distinctive twist in the posterior columella. The animal has a small head projecting forward beyond the shell and a short, tapering, curved tail extending posteriorly.

SIZE: shell length to 8 mm.

HABITAT AND RANGE: intertidal to 20 m depth; in shallow waters associated with the fine sediments upon which the filamentous green alga *Derbesia marina* grows. Vic. and Tas. to southern WA.

Cylichnatys campanula

The semitransparent shell is ovoid with parallel sides and bluntly rounded fore and aft ends. Its surface, viewed microscopically, is sculptured all over with fine, close, spiral **striae**. The animal is small and has the usual two posterior lobes on the headshield. The tail is short and rounded. The body is pale grey, sparsely to densely coloured with black, including on the mantle. A crimson sensory pad is present each side of the mouth.

SIZE OF SHELL: 3–4 mm long.

HABITAT AND RANGE: intertidal to 5 m depth; in fine muddy sand. NSW to southern WA, including Tas.

Austrocylichna exigua (Haminoeidae), Crawfish Rock, Western Port, Vic. Platon Vafiadis.

Cylichnatys campanula (Haminoeidae), Port Welshpool, Vic. Leon Altoff.

Haminoea maugeansis

Shell thin, smooth, glassy in life, and globular in shape. The animal is large when extended, about twice the length of the shell; the head is flattened and broad, with a pair of rounded lobes projecting posteriorly from the headshield visible dorsally and the eyes widely separated. The tail is short and rounded. The colour of the animal is usually dark, both externally and within the shell, with distinctive swirls of black intermingled with cream, yellow and orange splashes of pigment. Specimens from very exposed, sunny habitats, such as upper littoral pools, may be entirely black, with small orange spots as the only relief.

Size of shell: 12 mm, usually smaller.

Habitat and range: intertidal to 22 m depth; very common intertidally, crawling on muddy-sand sediments in and around seagrass beds and in shallow upper intertidal rock pools. Vic., Tas. and SA.

Liloa brevis

Shell thin, spirally striate towards each end, semitransparent, **elongate oval** in shape with parallel sides, sometimes covered with a pale-brownish periostracum. The animal is large when extended, almost twice the length of the shell; head flattened, with a pair of rounded lobes projecting posteriorly from the headshield. The tail is short and rounded. The animal is usually pale sandy fawn, often with sketchy greenish or greyish patches, and a few orange or red patches in the shell mantle. This species jerkily flaps its parapodia to move out of the path of predators.

Size of shell: to 20 mm, but usually much smaller.

Habitat and range: intertidal to 22 m depth; very common intertidally in protected bays, crawling through the muddy-sand sediments in and around seagrass beds. Qld to WA, including Tas.

Nipponatys tumidus

A very similar species to *Austrocylichna exigua*. The shell of this species is slightly more rounded, the posterior columella is less twisted, and the anterior part of the aperture is more oblique. The tail is very short, broad and obliquely truncate.

Size of shell: to 6 mm in length.

Habitat and range: intertidal to 20 m depth; in shallow water associated with the fine sediments upon which the filamentous green alga *Derbesia marina* grows. Vic. and Tas.

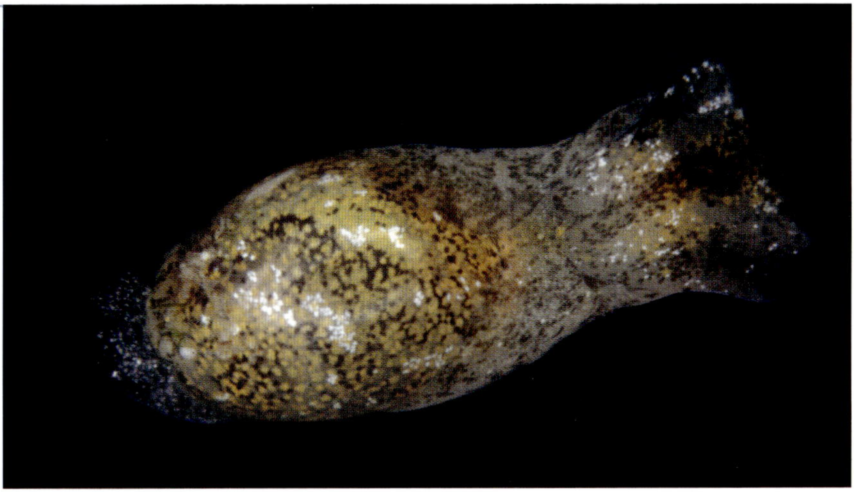

Haminoea maugeansis (Haminoeidae), Flat Rocks, Bunurong Marine Reserve, Vic. Platon Vafiadis.

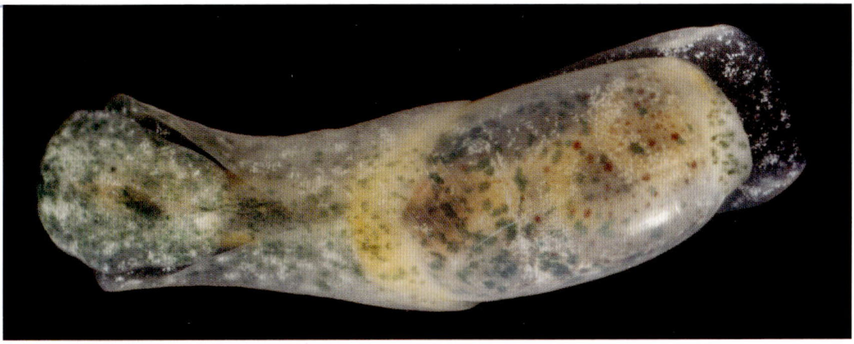

Liloa brevis (Haminoeidae), Clifton Springs, Port Phillip Bay, Vic. Leon Altoff.

Nipponatys tumidus (Haminoeidae), Crawfish Rock, Western Port, Vic.
Ventral view showing obliquely truncate tail. Platon Vafiadis.

Family Ilbiidae

A family of small sea slugs similar in size and shape to runcinids, presently known only from the western Pacific. The elongate or elongate-oval body has no transverse division into headshield and visceral mass. The foot is slightly longer than the body. A radula is present. Ilbiidae have neither gills, nor gastral plates, nor shell.

Ilbia ilbi

Readily recognised by its mauve or purple body colour ornamented with a pattern of yellow longitudinal lines on the mantle and tail.

Size: usually less than 5 mm.

Habitat and range: intertidal to 12 m depth; among seagrass and algae. Southern Qld, NSW and Vic.

Ilbia ilbi (Ilbiidae), Flat Rocks, Bunurong Marine Reserve, Vic. Platon Vafiadis.

Family Philinidae

Philinids are large-bodied animals enclosing a generally thin, fragile shell in the hinder part (the visceral mass or **visceral hump**). The shell has a very broad aperture and a very short (or no) spire. It may be smooth, except for incremental growth lines, or sculptured with many spiral rows of shallow but distinct grooves. The animal has thick lateral extensions (parapodia) of the foot folded up against the headshield and visceral mass, and has a radula with (usually two) large sickle-shaped teeth in each row, as well as three hard gastral plates in the stomach for crushing and grinding ingested food, often small **bivalves**.

Most philinids live in fine sandy or muddy habitats, and can be exceedingly common. A few are epifaunal species: these are usually small and sometimes strikingly coloured. Most philinids are creamy white, in keeping with their infaunal life. The egg mass of philinids is a balloon anchored by a thread to larger sand grains or shell fragments. The larger species lay flaccid, sausage-shaped balloons filled with strings of minute eggs.

There are numerous small and minute philinids in south-eastern Australian waters. Most are known only from deep water. Two unnamed species are included below.

Philine angasi

A common component of protected shallow-water embayments, sometimes confused with the naticid gastropod *Sinum zonale*, which has a similar body shape and colour, and a partly exposed solid shell. The body of *P. angasi* is entirely creamy white, elongate oval in outline, rather flattened, and very slippery to handle. In animals less than 20 mm long, the shell is very fragile and sculptured with shallow, spiral grooves, which are minutely **punctate** when viewed with a microscope. In larger adult animals, the shell is less fragile and completely smooth. The egg balloon is large, sausage-like, very soft and very conspicuous.

SIZE OF ANIMAL: 30–50 mm, exceptionally to 100 mm; size of shell: 15–30 mm.

HABITAT AND RANGE: intertidal to 500 m depth; crawling through the uppermost muddy-sandy sediment. All of southern Australia from Qld to WA; also NZ.

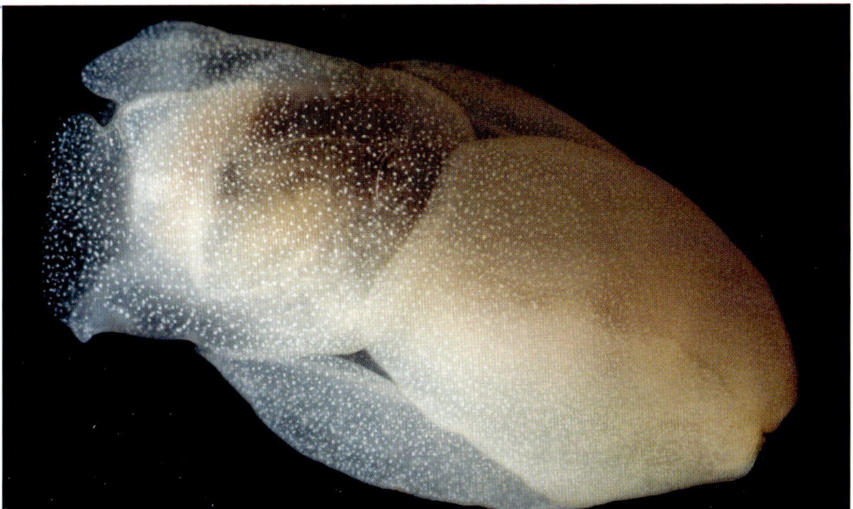

Philine angasi (Philinidae), Bastion Point, Mallacoota, Vic. Leon Altoff.

Philine sp. RB1

A small epifaunal species; headshield and visceral hump equal in length; orange in colour all over. Gastral plates small and rounded. Shell 3–4 mm long, **quadrately** rounded, sculptured with strong, spiral punctate grooves. Small size, body colour and epifaunal habitat separate *Philine* sp. RB1 from the larger infaunal *P. angasi*.

Size of animal: 6–9 mm.

Habitat and range: intertidal to about 50 m depth; on algae and seagrass in the lower intertidal zone. NSW to southern WA, including Tas.

Philine sp. RB2

A minute epifaunal species; elongate slender headshield; opaque white in colour with dense aggregations of white pigment cells on the visceral hump. Gastral plates absent. Shell hardly more than 1-mm long, rounded, sculptured with close spiral punctate grooves right to the aperture edge. Very small size and epifaunal habitat distinguish this species from the similarly coloured but much larger infaunal *Philine angasi*.

Size of animal: 2–3 mm.

Habitat and range: intertidal to 3 m depth; on algae in lower intertidal pools. Presently known only from Vic.

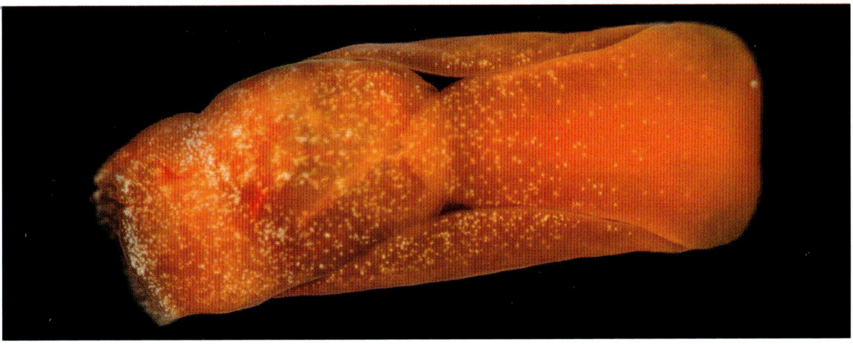

Philine sp. RB1 (Philinidae), Saxon Reef, Portland, Vic. Leon Altoff.

Philine sp. RB2 (Philinidae), Harmers Haven, Bunurong, Vic. Platon Vafiadis.

Family Retusidae

Most retusids are cylindrical with a depressed spire, very similar in shell shape to some smaller cylichnids. Some are very **pyriform**, others fusiform. A few retusids retain a small operculum, far smaller than the shell aperture into which the animal retracts. The retusid animal is small, one-quarter to one-third the shell length, with a pair of rounded lobes projecting from the rear edge of the head on to the anterior shell. All retusids lack a radula. The three gastral plates each have a pimpled, dimpled or crested inner face. Retusids can be especially abundant in intertidal and shallow water, mud and sandy mudflats.

Retusa atkinsoni

Shell long and slender, swollen in front, obliquely truncate behind; periostracum orange. Animal small and creamy white.

SIZE OF SHELL: 6 mm, usually smaller.

HABITAT AND RANGE: 0–100 m depth; fine sandy sediment. Southern Australia from NSW to SA.

Retusa pelyx

Shell more or less cylindrical, usually a little wider in front; periostracum when present yellowish or brownish; animal creamy white, the dark-brown pimpled gastral plates visible within the body behind the head; small, round reddish-brown operculum present in some specimens.

SIZE OF SHELL: 3–5 mm.

HABITAT AND RANGE: 0–100 m; in sandy mud, common in estuaries. Vic. and Tas.

There are at least six other species of *Retusa* in Victorian waters. Some are waiting on confirmation of identification; others are unnamed.

Retusa atkinsoni (Retusidae), Crawfish Rock, Western Port, Vic. Platon Vafiadis.

Retusa pelyx (Retusidae), Inverloch, Vic. Platon Vafiadis.

Family Runcinidae

A common family of molluscs related to the bubble shells, but because of their small size, they are rarely seen. The body is usually elongate oval, with no transverse division into headshield and visceral mass as present in the philinids, aglajids and gastropterids. The foot is as long or longer than the body. Gills are present between the foot and the mantle, either at the rear right of the body or in the posterior midline. There is usually no shell, but some have a minute, internal shelly plate posteriorly, and at least two species have a minute, external coiled shell projecting from a notch in the posterior mantle edge. Runcinids have a radula and four transversely ribbed gastral plates.

Runcina australis

This species is black, with somewhat lighter anterior corners to the head. Its gills are on the posterior right of the body. A minute, whitish shelly plate is sometimes present.

Size: usually less than 5 mm, exceptionally to 7 mm.

Habitat and range: intertidal to 10 m depth; often on the wire weed, *Amphibolis antarctica*, or crawling on algae in shallow, lower intertidal pools. NSW, Vic. and SA.

Runcina sp. RB1

A broader, stouter animal than *Runcina australis*, with a minute, partly coiled external shell projecting from the notch in the rear edge of the mantle. The gills are on the posterior right of the body. Body colour is dark brownish, with a whitish transverse bar across the mantle at about the anterior third, and traces of another at the third quarter. Eyes are visible in semitransparent, teardrop patches near the front edge.

Size: 2–4 mm.

Habitat and range: 0–5 m depth; intertidally associated with brown algae, corallines, the seagrass *Amphibolis* and the filamentous turfing green alga *Derbesia marina*. NSW and Vic.

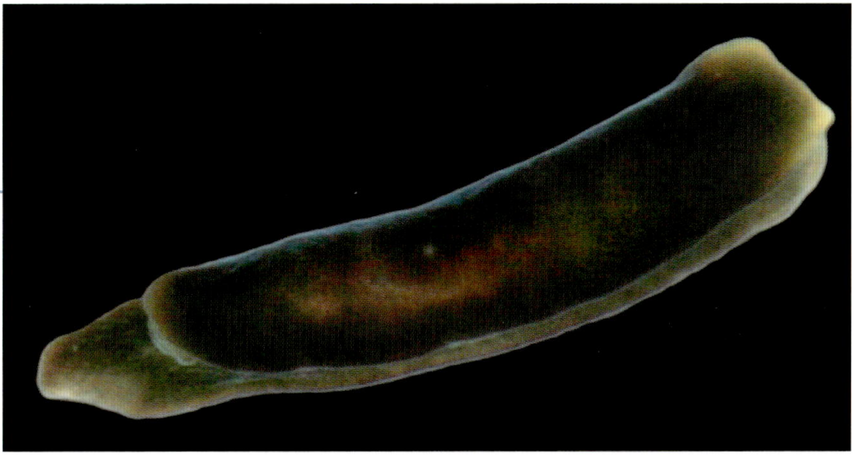

Runcina australis (Runcinidae), Anderson Point, Portland, Vic. Platon Vafiadis.

Runcina sp. RB1 (Runcinidae), East Cape, Cape Conran, Vic. Leon Altoff.

Family Volvulellidae

This recently separated family has a considerable number of species, most of which occur in subtidal to very deep waters. Very little is known of the animal, its habitats and habits. The shell is spindle-shaped, with a long aperture extending the whole length of the shell. In some species, the posterior end of the aperture narrows and tapers to a very sharp point, while in others it is shorter and obliquely truncate. The animal is slender, the head with a pair of tapering strap-like lobes that do not reach the shell. There are neither radula nor gastral plates in this family.

Volvulella rostrata

Shell spindle-shaped, narrowly rounded in front, obliquely pointed behind. Shell and animal white, head parts almost half length of shell, eyes prominent.

Size of shell: 5–7 mm.

Habitat and range: 0–100 m; southern Australia from NSW to WA. Rarely found alive.

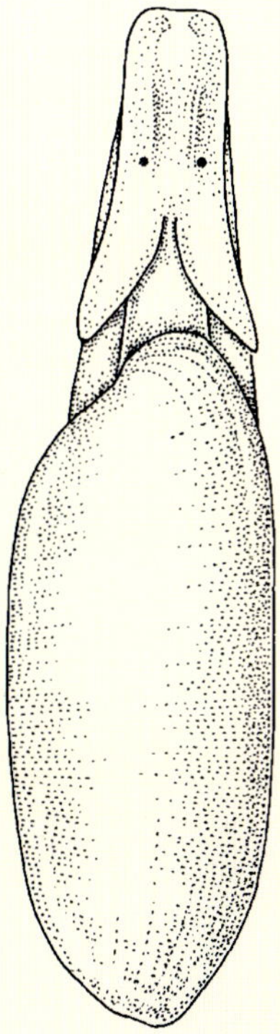

Volvulella rostrata (Volvulellidae), Oyster Harbour, WA. Illustration: R. Plant from Beesley, Ross and Wells (1998) (no live image available).

Order Sacoglossa

These are the sap-sucking sea slugs, characterised by buccal morphology that allows them to pierce the walls of certain green and red algae with their knife-like radular teeth and suck out the contained fluids. Many species retain living chloroplasts in the fine branching of the digestive gland within their body and cerata. The fossil record indicates that sacoglossans have existed at least from the Eocene. The order is diverse in external form, from species with an external shell into which the animal can retract, to naked species with parapodia or cerata along each side of the body. A further character, useful in separating sacoglossans from similar-looking nudibranchs, is the presence of only two tentacles (rhinophores) rising from the head.

Southern Australia is rich in sacoglossan genera and species, many of which remain unnamed.

Polybranchia pallens

Body broad and rather flat, cerata with short, thick, cylindrical stalk and broadly fan-shaped upper lobe; digestive gland with its **dendritic** branching visible within lobe; margin of lobe with a continuous series of small points. Body cream, medially pink; cerata usually fawn or greenish brown, sometimes with a darker spot in the middle of each lobe.

Size of animal: 40 mm, but usually smaller. Wholly pale-pink specimens 5–7-mm long are newly hatched juveniles.

Habitat and range: intertidal to 10 m depth; often found near *Caulerpa flexuosa* in rock pools, but no association confirmed. Vic., Tas. and eastern SA.

Family Caliphyllidae

Caliphyllids are stout aeolidiform sacoglossans with leaf-like cerata along each side of the body. The rhinophores are each deeply **bifid** (forked), giving the impression of four, rather than the two head tentacles definitive of the order.

Polybranchia pallens (Caliphyllidae), West Head, Flinders, Vic. John Chuk.

Family Hermaeidae

The cerata of hermaeids are usually fusiform and often almost transparent except for the digestive gland duct and its branches. A few species have paddle-shaped cerata, but these are smooth edged and thick. The rhinophores are laterally grooved, slightly wider at the tip, each with two apical lobes; some hermaeids have smooth, simple rhinophores. Some species are brilliant (even fluorescent) red in colour in life, acquiring their pigmentation from the red algae they eat.

In the checklist of Burn (2006), members of the Hermaeidae were inadvertently treated as belonging to the Limapontiidae; here Burn (1998) is followed, and the family is maintained as distinct. There are at least five unnamed species of Hermaeidae known from the Bass Strait area.

Aplysiopsis formosa

Animal long and slender; rhinophores short, cylindrical and laterally grooved; cerata fusiform, long and slender, with the medial side showing the ridge-like impression of a slender, branching tree. Body dark green; a large triangular black patch on the head, black stripes on the cerata, and two broad, black parallel lines on the sole of the foot.

Size of animal: 30 mm, seldom smaller than 10 mm.

Habitat and range: intertidal to 15 m depth; found with the alga *Cladophora prolifera*, especially in deep sand-filled rock pools. Central Vic.; temperate North Atlantic and Mediterranean.

Hermaea evelinemarcusae

The pigmented internal organs and ducts can be readily seen in this transparent species. The slender body tapers to a long thin tail. The head is a little wider, with short apically swollen and laterally grooved rhinophores. The long, slender cerata are irregularly fusiform, the brown digestive gland ducts within being sparsely branched. The body and cerata have a reddish tinge from the minute reddish-brown specks that cover the entire animal. The rhinophores have a whitish band below the tip, and the alimentary and reproductive masses show as white and cream within the body.

Size: 8–10 mm.

Habitat and range: 0–5 m depth; usually on epiphytic algae on seagrasses *Zostera* and *Amphibolis*. Central NSW, Vic., SA and southern WA.

Aplysiopsis formosa (Hermaeidae), Point Lonsdale, Vic. Platon Vafiadis.

Hermaea evelinemarcusae (Hermaeidae), Balnarring Beach, Vic. John Chuk.

Hermaea sp. RB1

The cerata of this species are long and paddle-shaped, with the red, pink or orange digestive gland branching in one plane only and almost entirely filling each ceras. The body can be white, pink or red, almost always with a white patch on the pericardial hump between the anterior cerata.

Size: 8–15 mm, exceptionally to 25 mm body length, with cerata 10–12 mm high.

Habitat and range: 0–13 m depth; often associated with red algae, from which it derives both nutrition and colour. NSW to SA.

Family Juliidae

The bivalved gastropods. The shell in these sacoglossans commences as a typical spiral embryonic protoconch, but once the adult shell (**teleoconch**) starts to form, anterior and posterior indentations in the depositing or growing edge of the mantle create a left 'valve' (retaining the protoconch at the apex or **umbo** and containing most of the visceral mass) and a slightly smaller, right 'valve' (containing the organs of the mantle cavity). The animal is slender, the tail short, the eyes are distinct and close together in a small clear dome on the top of the neck, and the rhinophores are short and cylindrical. The animal can withdraw completely within the shell and close the 'valves' tightly together. Discharges a white viscid fluid when disturbed. Bivalved gastropods are very **cryptic** on their *Caulerpa* host-food. Four species of Juliidae are presently known from south-eastern Australia, three of which are illustrated here. Some authors treat these as belonging to the genus *Berthelinia*.

Edenttellina typica

Shell elongate oval when viewed from the side, protoconch flat. Animal pale yellowish green; shell mantle with several more or less horizontal dark stripes.

Size of animal: 5 mm; size of shell: 4 mm.

Habitat and range: intertidal to 2 m depth; obligate on *Caulerpa brownii*. Central Vic., northern Tas. and eastern SA.

Hermaea sp. RB1 (Hermaeidae), Portsea Pier, Vic. Mary Molloy.

Edenttellina typica (Juliidae), Point Hawdon, Kennett River, Vic. Platon Vafiadis.

Midorigai australis

Shell high in front with steep, posterior slope when viewed from the side; protoconch flat. Animal dark green; shell mantles with numerous radials of creamy-yellow spots, increasing in size from the umbo.

Size of animal: 6 mm; size of shell: 5 mm.

Habitat and range: intertidal to 2 m depth; obligate on *Caulerpa simpliciuscula*. Central Vic., northern Tas., SA and southern WA.

Tamanovalva babai

Shell more or less triangular when viewed from the side, with steep slope from front to rear, protoconch erect. Animal green or pale green; shell mantle green or pale green, often with small, cream marginal patches.

Size of animal: 10 mm; size of shell: 8 mm.

Habitat and range: intertidal to 3 m depth; crawling and feeding on *Caulerpa*, especially species with wide smooth surfaces. Central Vic., northern Tas. and eastern SA.

Midorigai australis (Juliidae), Kitty Miller Bay, Phillip Island, Vic. Platon Vafiadis.

Tamanovalva babai (Juliidae), Kitty Miller Bay, Phillip Island, Vic. Platon Vafiadis.

Family Limapontiidae

Mostly small, aeolidiform species with fusiform cerata and simple rhinophores, though there are a few species without either cerata or rhinophores. Differences in the shape of the radular teeth and the reproductive systems separate limapontiids from other aeolidiform sacoglossan families.

There are numerous unnamed limapontiids known from the Bass Strait area.

Costasiella sp. RB1

Very similar to the following species in shape and colour, but is separated by the domed or convex **pericardium** and the wide, tentacle-like anterior foot corners.

Size: 6–8 mm.

Habitat and range: 0–5 m depth; associated with the filamentous green alga *Derbesia marina*. Southern NSW to central Vic.

Costasiella sp. RB2

Eyes close together between the rhinophoral bases distinguish this genus. A drab little species: yellowish-green to dark brownish-green cerata on a dirty white body with a short tail. The rhinophores are long, tapering and held at right angles to the body. A black line on the anterior edge of the rhinophores continues forward and marks the edges of the head. The pericardium is large and dorsally depressed, forming a tray-like structure between the cerata. The anterior foot is broad and rounded.

Size: 4–6 mm.

Habitat and range: lower intertidal; associated with the filamentous green alga *Derbesia marina*. Central Vic.

Costasiella sp. RB2 (Limapontiidae), Shallow Inlet, Vic. Leon Altoff.

Ercolania boodleae

Slightly larger and stouter than the following species, with long, tapering rhinophores, flat anal aperture on the front right of the pericardium, and crowded (20–30) fusiform cerata along each side of the body. Animal entirely shining black except for the orange or red tips to the cerata and long, bright white cheeks at the lower base and extending up the rhinophores.

Size of animal: 12 mm, often smaller.

Habitat and range: intertidal to 6 m depth; loosely associated with a range of filamentous and harder green algae. Apparently introduced to Port Phillip Bay and Western Port, Vic., where first recorded in 2002. Native range: Japan and southern China.

Ercolania margaritae

A small species with short, cylindrical rhinophores; high anal **papilla** at the rear right of the pericardium, and rather few (12–15) stout cerata each side of the body. Intertidal rock pool animals are often entirely dull black with only paler tips to cerata and cream cheeks at the lower base of the rhinophores. Subtidal animals (such as the one illustrated here) are paler, the cerata often suffused with maroon or green, with only the tips blackish.

Size of animal: 8–10 mm, usually smaller.

Habitat and range: intertidal to 16 m depth; often found on fine filamentous green algae such as *Enteromorpha* and *Cladophora*. Southern NSW to southern Tas. to western Vic. Can be extremely common on intertidal rock platforms, but the colouration tends to make the animals very inconspicuous.

Ercolania sp. RB2

Another small species, mottled brown in colour, usually with darker head and lighter foot. Body stouter than *Ercolania margaritae*, the rhinophores very short and little more than elevated crests, and anal papilla projecting from the right of the **pericardial ridge** near the end of the body.

Size: 8–10 mm, usually smaller.

Habitat and range: lower intertidal; associated with various filamentous green algae. Southern NSW to central Vic.

Ercolania boodleae (Limapontiidae), Flinders Pier, Vic. John Chuk.

Ercolania margaritae (Limapontiidae), Kitty Miller Bay, Phillip Island, Vic. Platon Vafiadis.

Ercolania sp. RB2 (Limapontiidae), Clifton Springs, Vic. Platon Vafiadis.

Placida dendritica

There are several species of *Placida* in the Bass Strait area, differing in food choice, size, colours and shape of cerata. *Placida dendritica* is an all-greenish-yellow species, throughout which the dark-green digestive gland branches are visible. The rhinophores are cylindrical and shallowly grooved along one side. The cerata are fusiform and, besides the branching digestive gland, contain yellowish branches of the albumen gland.

Size of animal: 8–10 mm, often smaller.

Habitat and range: intertidal to 10 m depth; associated with the green alga *Codium fragile*, upon which it lives and feeds. All Australian states; circumglobal tropical and temperate seas.

Stiliger smaragdinus

The largest sacoglossan in the Bass Strait area, often seen crawling over algae in deep intertidal rock pools. Animal pale green to bluish green with a distinctive cream and black bar across the rhinophoral bases. Rhinophores large, deeply grooved along outer side. Cerata numerous and crowded, almost covering the entire **dorsum** (upper mantle surface). Each ceras has a stout stalk and globular top, and on the inner side of each stalk is a distinctive thick flange, below which is a circlet of small, white glands that discharge a white viscid fluid upon disturbance or handling.

Size of animal: 100 mm, but usually much smaller.

Habitat and range: intertidal to 10 m depth; associated with the green alga *Caulerpa* (especially those species with globular vesicles: *C. cactoides, C. geminata, C. racemosa* and *C. vesiculifera*), which the animal closely resembles both in colour and disposition of the cerata. All Australian states; NZ, tropical and temperate Indo-Pacific.

Placida dendritica (Limapontiidae), Gabo Island, Vic. Leon Altoff.

Stiliger smaragdinus (Limapontiidae), Kitty Miller Bay, Phillip Island, Vic. Platon Vafiadis.

Family Oxynoidae

Animal larger than shell, often with long tail; two long to short parapodial lobes either side of body next to the shell. Shell rounded, thin and flexible, with large aperture. Discharges a white viscid fluid when disturbed.

Oxynoe viridis

Shell yellowish, **bulliform**, with very wide aperture. Animal slender and much longer than shell; rhinophores long and cylindrical; short parapodia with several bigger and smaller conical papillae on the outer side, lying on shell; tail more than half the length of animal. Animal greenish to yellowish green, with dense-to-sparse yellow, black and blue spotting all over; rhinophores often with blue-and-white banding. Tail may break off when handled.

Size of animal: 30 mm; size of shell: 10 mm, but both usually smaller.

Habitat and range: intertidal to 10 m depth; crawling and feeding on and among various green algae, but mostly *Caulerpa*. All Australian states; tropical and temperate Indo-Pacific.

Roburnella wilsoni

Shell yellowish, bulliform, with wide aperture. Animal larger than shell; rhinophores long and cylindrical; parapodia long, narrow, slightly rolled, smooth and often standing away from shell; tail short and broad. Animal pale green with cream flecks and network all over; parapodia with blue-and-orange bands at tip.

Size of animal: 15 mm; size of shell: 10 mm, but both usually smaller.

Habitat and range: intertidal to 22 m depth; crawling and feeding on *Caulerpa*, especially species with wide smooth surfaces. Central Vic. and northern Tas. to southern WA.

Oxynoe viridis (Oxynoidae), Gibson Steps, western Vic. Leon Altoff.

Roburnella wilsoni (Oxynoidae), Point Lonsdale, Vic. Platon Vafiadis.

Family Plakobranchidae

Animals without shell, except during early development. Body usually slender and soft, with a parapodial flange along each side of the posterior two-thirds of the body length. The raised pericardium lies between the anterior ends of the parapodia. The pattern of the raised ribs leading to the pericardium can be useful in separating those species that have them.

Seven species are currently known from the Bass Strait region, of which several remain unnamed. Six species are illustrated here.

Elysia coodgeensis

Small, slender animal, usually pale green in colour but may be blackish or brownish green, always with a short, black stripe between the rhinophores, a narrow, black line along each parapodial margin, bluish-black tips to the rhinophores and tail, and a few globular, cream glands submarginally within the parapodia.

Size of animal: 20 mm, but usually smaller.

Habitat and range: intertidal to 3 m depth; associated with filamentous green algae, such as *Enteromorpha* and *Chaetomorpha,* in intertidal rock pools and on sand flats. Central and eastern Vic.; all Australian states except Tas. Previously misidentified as *Elysia australis.*

Elysia furvacauda

Small, very slender animal, usually reddish brown in colour but may be dark green, everywhere speckled with cream, yellow, orange, blue and black. Tips of rhinophores and tail blackish; parapodial margins when held together form two exhalent holes, the margins being cream at the openings.

Size of animal: 25 mm, but usually smaller.

Habitat and range: intertidal to 3 m depth; possibly associated with the clumping green alga *Cladophora.* Confirmed distribution is central Vic. to central SA.

Elysia coodgeensis (Plakobranchidae), Inverloch, Vic. T. Joan Hales.

Elysia furvacauda (Plakobranchidae), Anderson Point, Portland, Vic. Platon Vafiadis.

Elysia maoria

Small slender animal, usually dark green in colour with a scattering of minute black and yellow specks; tips of rhinophores and tail darker.

SIZE OF ANIMAL: 15 mm, often smaller.

HABITAT AND RANGE: intertidal to 4 m depth; associated with species of the green alga genus *Codium*. Southern Qld to western Vic.; NZ.

Elysia sp. RB1

Separated from other Victorian species by the two, usually long, tongue-like lobes projecting from each parapodial margin. Big specimens may have a smaller third lobe posterior to the second pair. Body slender and soft; rhinophores cylindrical. Body colour usually reddish brown, but may be darker or even greenish.

SIZE: 20 mm, usually smaller.

HABITAT AND RANGE: 0–10 m depth; crawling on algae. Vic.

Elysia sp. RB2

Though similar in colour to *Elysia coodgeensis*, this species has each parapodial margin raised into two black shark-fin projections, sometimes with indications of a third projection. The rhinophores are long, tapering and black tipped; the tail is long and slender.

SIZE: 6–10 mm.

HABITAT AND RANGE: lower intertidal and shallow subtidal; associated with the green alga *Caulerpa cactoides* growing in silty conditions. Central Vic.

Elysia maoria (Plakobranchidae), Portsea Pier, Vic. John Chuk.

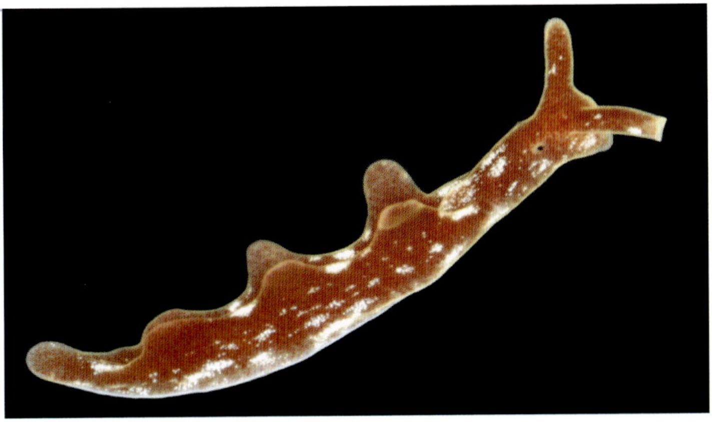

Elysia sp. RB1 (Plakobranchidae), Sunderland Bay, Phillip Island, Vic. Platon Vafiadis.

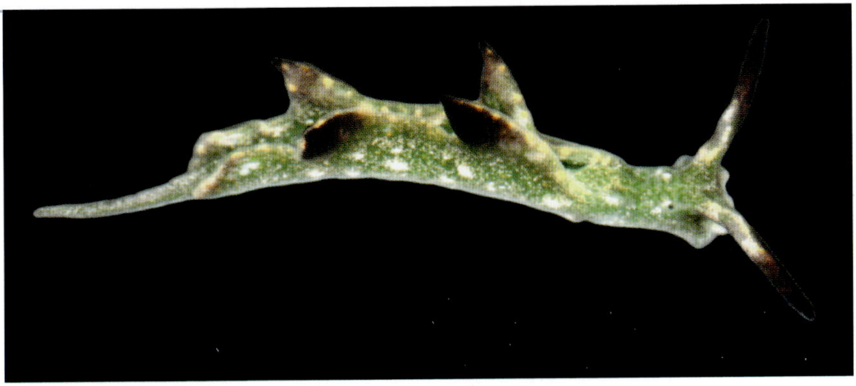

Elysia sp. RB2 (Plakobranchidae), Crawfish Rock, Western Port, Vic. Platon Vafiadis.

Elysia sp. RB3

This species resembles a stouter version of *Elysia furvacauda*, with higher, smooth-margined parapodia and longer, cylindrical rhinophores. Body colour varies from reddish brown to brownish green; rhinophoral and tail tips bluish black.

Size: 10–15 mm.

Habitat and range: lower intertidal, in areas close to *Zostera muelleri* beds. Central Vic. and SA.

Family Volvatellidae

Shell cylindrical or pear shaped with posterior spout; thin, flexible, more or less chitinous with little shelly component. Animal small with a short foot; capable of retracting into the shell; gives off a white viscid discharge when disturbed.

Ascobulla fischeri

Shell cylindrical, posteriorly dome shaped with a spiral slit along the suture; yellowish. Animal creamy yellow; head with a pair of pad-like, rounded lobes in front and behind.

Size of animal: 15 mm; size of shell: 12 mm.

Habitat and range: intertidal to 15 m depth; always associated with the root system of green algae of the genus *Caulerpa*. All of southern Australia, from Qld to southern WA.

Elysia sp. RB3 (Plakobranchidae), Harmers Haven, Vic. T. Joan Hales.

Ascobulla fischeri (Volvatellidae), Sunderland Bay, Phillip Island, Vic. Platon Vafiadis.

Order Anaspidea

Sea hares (another group of shallow-water herbivorous sea slugs). Here the majority are large species, including the largest, *Aplysia vaccaria*, from California and Mexico, which attains 900 mm in length and almost 14 kg in weight. Sea hares usually have a narrower head-end with forward-projecting oral tentacles and dorsally directed rhinophores shaped rather like rabbits' ears. The visceral mass makes up the bulky posterior part of the animal, behind which extends a short or long tail. All sea hares have parapodia, in some species well developed and standing upwards, in others small and folded close to the body to protect the mantle cavity. There is an external shell enveloping the visceral mass in the family Akeridae. In the family Aplysiidae, the shell is either absent, or reduced and internal, sitting over the visceral mass. Some species in each family can swim using their parapodia.

Akera soluta

Shell pale brown, animal darker brownish.

SIZE OF ANIMAL: 30 mm; size of shell: 20 mm.

HABITAT AND RANGE: intertidal to 10 m depth; crawling and burrowing in fine sediment near seagrass beds. Estuaries and embayments of eastern Australia; tropical and temperate Indo-Pacific. Sometimes present in vast numbers.

Family Akeridae

Shell external, fragile, flexible and stoutly cylindrical with a narrow **sutural slit**. Animal larger than shell; head without oral tentacles and rhinophores; parapodia large (folding over anterior part of shell to prevent sediment entering the mantle cavity while burrowing), used for swimming. The species included here occurs just outside the geographic area of this book, but it may at times be found in the estuaries of far-eastern Victoria.

Akera soluta (Akeridae), Koumac, New Caledonia. Bill Rudman, Australian Museum.

Family Aplysiidae

Shell internal or absent. Temperate Australian species with a narrower head-end and a bulbous visceral mass posteriorly; they have smooth, tuberculate or papillate bodies and discharge pigmented repugnant fluids upon disturbance or handling. None of the species listed herein swim. Additional species occur in the Bass Strait area, but their identities have yet to be confirmed.

Aplysia dactylomela

Pale yellowy-brown animal with large black rings scattered across the body and parapodia; medial side of parapodia with large patches of black and fawn; tail tip often black; discharges reddish-purple fluid. Recent molecular phylogenetic studies (2013) indicate that *Aplysia dactylomela* is restricted to Atlantic waters, and that Indo-Pacific specimens should be identified as *Aplysia argus*.

Size of animal: to 180 mm.

Habitat and range: intertidal to 20 m depth; intertidal rock pools. Circumglobal tropical and temperate seas. Uncommon in Bass Strait area.

Aplysia juliana

Body somewhat bulky, parapodia fused high up at rear; foot has the sole divided transversely into two sections, which allows animal to crawl along in an inching fashion; animal uniformly greyish brown with medial sides of parapodia darker, small but distinct white patch around each eye; discharges whitish fluid.

Size of animal: up to 200 mm.

Habitat and range: intertidal to 20 m depth; intertidal rock pools and seagrass beds. Circumglobal tropical and temperate seas.

Aplysia dactylomela (Aplysiidae), Sydney, Long Reef, NSW. Bill Rudman, Australian Museum.

Aplysia juliana (Aplysiidae), Mallacoota Inlet, off east side of Goat Island, Vic. Leon Altoff.

Aplysia parvula

Intertidal animals are brownish or brownish red; subtidal animals are invariably deep red; parapodial margins, tail and rhinophoral tips always black; discharges reddish-purple fluid.

Size of animal: usually less than 50 mm, exceptionally up to 100 mm.

Habitat and range: intertidal to 50 m depth; among algae and seagrass. Circumglobal tropical and temperate seas.

Aplysia sowerbyi

Another larger species, identified by the **anastomosing** vein-like black lines forming a network on the outside of the parapodia. The discharge fluid is reddish purple.

Size: 180 mm, usually smaller.

Habitat and range: 0–20 m; among algae and seagrass. NT, Qld, NSW and Vic.

Aplysia sydneyensis

Animal small with large parapodia, body slender, somewhat transparent with the reddish pharynx obvious within the head; pale brownish patches and their brown vein-like lines mark the body and parapodia; discharge dull reddish.

Size of animal: up to 100 mm, usually smaller.

Habitat and range: intertidal to 10 m depth; among algae and seagrass. NSW and Vic.

Aplysia parvula (Aplysiidae), Port Campbell Harbour, Vic. John Chuk.

Aplysia sowerbyi (Aplysiidae), West Cape, Cape Conran, Vic. Leon Altoff.

Aplysia sydneyensis (Aplysiidae), Inverloch, Vic. T. Joan Hales.

Bursatella sp. RB1

Separated from typical *Bursatella leachii* by the dense coat of short, slender, pointed papillae, the long, narrow neck, and the humped (rather than swollen) body. Overall body colour dark brown.

Size: to 100 mm.

Habitat and range: 2–5 m depth; among algae and seagrass. Southern WA (where commonest), SA and rarely to central Vic.

Dolabrifera brazieri

A large, flattened species with low, fused parapodia, very large foot extending to all ventral margins, narrower head with prominent **enrolled** oral tentacles and smaller rhinophores; whole dorsal surface with extendable, compound, pointed papillae; small, triangular shell present in mantle cavity. Body colour variable—dorsally greenish or brownish but with patches, bands or stripes of darker or lighter colours; sole yellowish green, with a brighter green margin, and without any darker speckling.

Size of animal: up to 150 mm.

Habitat and range: intertidal to 5 m depth; in rock pools among algae and corallines. Southern Qld to eastern Vic.; NZ.

Bursatella sp. RB1 (Aplysiidae), Blairgowrie Marina, Vic. Nicola Patron.

Dolabrifera brazieri (Aplysiidae), Cape Conran, Vic. John Eichler.

Order Umbraculida

Umbrella shells. The few species in this order have a low, limpet-like shell perched on top of the visceral mass, and a large pinnate gill projecting into the right-hand space between shell and foot. All the species are grazers on sponge tissue.

Tylodina corticalis

Periostracum extending well beyond shell margin, broad, radial periostracal ribs present. Animal much larger than shell; tail rounded.

Size of animal: 40 mm, but usually smaller.

Habitat and range: intertidal to 100 m depth; feeds on sponges of the same yellow colour, often observed in deep rock pools in April–May each year. Southern Australia, from southern Qld to mid-WA.

Family Umbraculidae

Shell thick, solid and flattened, with thin periostracum present in places, often with attached massive growths. Animal large, high, tuberculate all over, with short tail and deep anterior groove, wherein lie the rhinophores, oral tentacles and penis. The gill fills the anterior and right-hand cavity between shell and foot.

Umbraculum umbraculum

Shell creamy white, with ventral yellow or orange central area. Animal big and bulky, thick and solid, varying in colour from creamy yellow to dark brown; dorsal surface tuberculose, often with the tubercles highlighted in a different colour.

Size of animal: up to 150 mm; size of shell: 100 mm.

Habitat and range: intertidal to 100 m depth; associated with the sponges upon which it feeds. Circumglobal tropical and temperate seas; trawled in deep water in eastern Bass Strait.

Family Tylodinidae

Shell thin and fragile, covered with a thick, ribbed or thatch-like periostracum. Animal large, elongate and smooth, with a long, rounded tail and large, basally swollen, laterally grooved rhinophores, enrolled oral tentacles, gill on right side; invariably bright yellow in life.

Tylodina corticalis (Tylodinidae), Harmers Haven, Bunurong Marine Reserve, Vic. Platon Vafiadis.

Umbraculum umbraculum (Umbraculidae), Lord Howe Island, NSW. Ian Hutton.

Order Pleurobranchida

Side-gilled slugs. Soft-bodied, oval animals, usually rather flattened, the mantle often smooth but in some larger species covered with many rounded or polygonal tubercles. Basally joined, enrolled rhinophores project from under the anterior mantle edge. In some species, the rhinophores are widely separated and rise from the dorsal surface. A broad velum, with short tentacle-like corners, is present across the head in all species. A feathery gill is always present along the right body wall between mantle and foot. A thin, fragile shell (or the periostracal remains) may be present within a reduced shell cavity above the visceral mass. Pleurobranchs feed on sponges or ascidians, or are omnivorous.

Berthellina citrina

Shell narrowly triangular and fawn; animal orange or deep yellow, sometimes with a few paler patches on mantle.

Size of animal: up to 35 mm; size of shell: 6 mm.

Habitat and range: intertidal to 150 m depth. Indo-Pacific tropical and temperate seas; not uncommon intertidally in central and eastern Vic.

Berthella medietas

Shell broadly oval and pale brownish; animal cream, yellow or pale fawn, often with a few whitish flecks and patches. Dorsal mantle honeycombed with clear glands.

Size of animal: up to 40 mm; size of shell: 10 mm.

Habitat and range: intertidal to 50 m depth; associated with the thin, slimy sponge *Plakina* under low intertidal and subtidal stones. Southern NSW to southern WA; NZ.

Family Pleurobranchidae

Description as for the order Pleurobranchida.

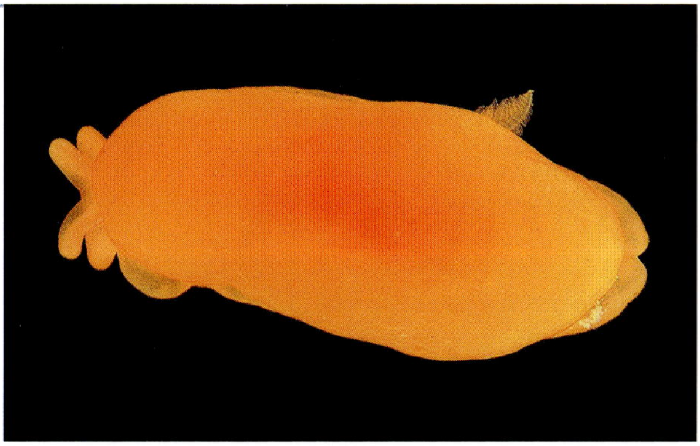

Berthellina citrina (Pleurobranchidae), San Remo, Vic. T. Joan Hales.

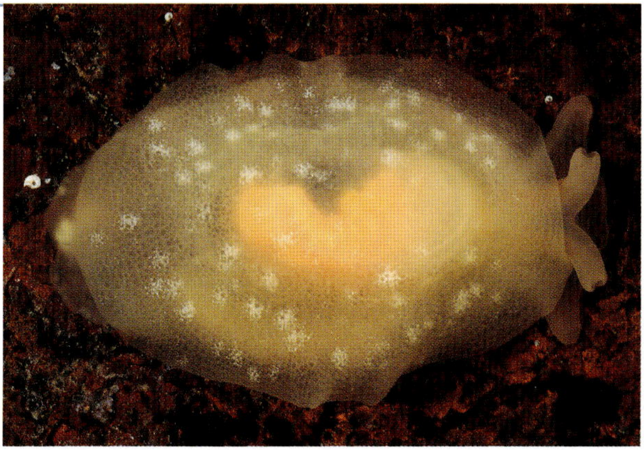

Berthella medietas (Pleurobranchidae), Point Leo, Western Port, Vic. John Chuk.

Berthella serenitas

Shell narrowly ovate, white with brown protoconch; animal transparent white, sometimes with dark spot over shell protoconch, dark mantle margin and rhinophoral tips.

Size of animal: up to 20 mm; size of shell: 6 mm.

Habitat and range: intertidal; under stones, associated with the slimy sponge *Plakina*. Known only from Vic.

Pleurobranchaea maculata

Animal elongate with mantle smaller than foot, gill exposed along right-hand body wall, velum broad, rhinophores widely separated; body colour fawn or greyish, sometimes with lighter or darker dappling. There is no shell.

Size of animal: 60 mm, but usually smaller.

Habitat and range: intertidal to 50 m depth; common under stones in intertidal rock pools. Southernmost Qld to WA; NZ.

Pleurobranchus hilli

A large rounded species with tuberculate mantle and a distinctive anterior notch through which the rhinophores project; mantle much larger than foot. There is no shell. Two colour forms occur and may be found together: wine red or maroon, or dull yellowish orange with a few larger tubercles encircled with maroon.

Size of animal: up to 400 mm, but usually less than 200 mm.

Habitat and range: low intertidal to 46 m depth; on sandy substrate associated with the ascidians upon which it feeds. Central NSW to SA; northern NZ.

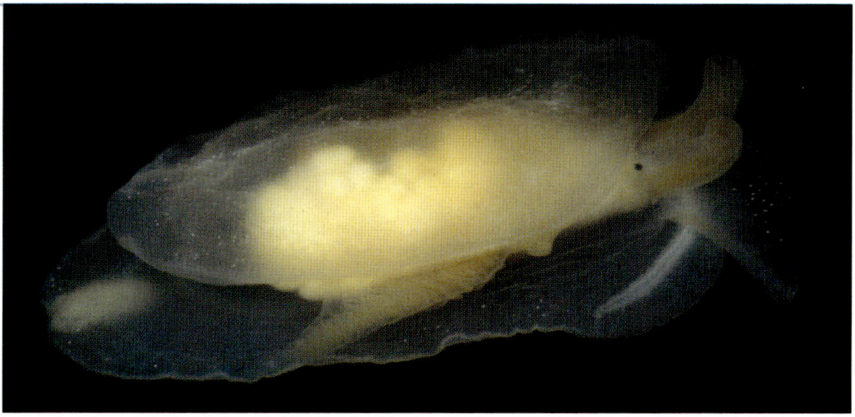

Berthella serenitas (Pleurobranchidae), Flat Rocks, Bunurong Marine Reserve, Vic. Platon Vafiadis.

Pleurobranchaea maculata (Pleurobranchidae), Barwon Heads, Vic. Platon Vafiadis.

Pleurobranchus hilli (Pleurobranchidae), Portsea Pier, Vic. John Chuk.

Order Nudibranchia

These are the true nudibranchs. Diverse in shape, size and colour, these animals entirely lack a shell, except briefly during embryonic development. Many species have minute, glassy, calcareous spicules reinforcing the dorsal epithelium; some few (Onchidorididae) have a dense mat of spicules visible to the naked eye within the mantle. The gills of nudibranchs are wholly exposed, usually as a cluster of feather-like plumes projecting from the posterior part of the dorsal mantle, as tree-like processes along each lateral edge of the mantle or body, or as globular or spindle-shaped cerata along each side of the body. A few species even have gill lamellae projecting from the body wall posteriorly (*Corambe*) or laterally (Arminidae) in the groove between mantle and foot. Nudibranchs are carnivores, their foodstuffs including sponges, ascidians, many and varied cnidarians, bryozoans, kamptozoans and other nudibranchs (even their own kind).

Southern Australia is rich in nudibranch genera and species, and many, though common, remain unnamed.

Family Actinocyclidae

Soft, highly convex dorids with a circle of upstanding gills and a velvet-like mantle surface. Peculiarities of the hard parts of the alimentary tract set this family apart from other dorids. There are several large, tropical Indo-Pacific species. However, most are small and uniformly coloured.

Hallaxa michaeli

A small species of uniformly pale-cream to pale-yellow colouration, somewhat translucent, the rhinophores bulbous at the upper half, and the gills forming a neat upstanding circle.

Size: 15 mm, but often smaller.

Habitat and range: intertidal to 55 mm depth; often found crawling on brown algae in rock pools. NSW to Vic. and Tas.

Suborder Doridina

Typical dorid nudibranchs with either a broadly oval notum projecting beyond the body, or the remnants of the projecting mantle in the form of smooth or dendritic papillae placed along the **pallial line** on each side of the body and across the head. A pair of chemosensory rhinophores stand near the anterior end of the mantle and a circlet of gills (**branchiae**) surrounding the anus project from the centreline near the posterior end of the mantle. Many species can withdraw rhinophores and gills into protective cavities within the mantle. A few species, all small, lack gills; one south-eastern Australian species (*Corambe* sp. RB1) has a few gills between the posterior foot and mantle, while the tropical family Phyllidiidae has a band of gill lamellae along each side of the body between mantle and foot. All doridinans are carnivores, using their radular teeth to scrape their food into the gut, or pharyngeal enzymes to externally turn their sponge food into 'soup' to slurp-up and pump to the stomach. Sponges, ascidians, bryozoans and kamptozoans are favoured foods, and many species are very cryptic when upon their food. Some species can swim by mantle undulation, others by vigorously flexing the body. Most, however, can do no more than crawl sedately over food and substrate.

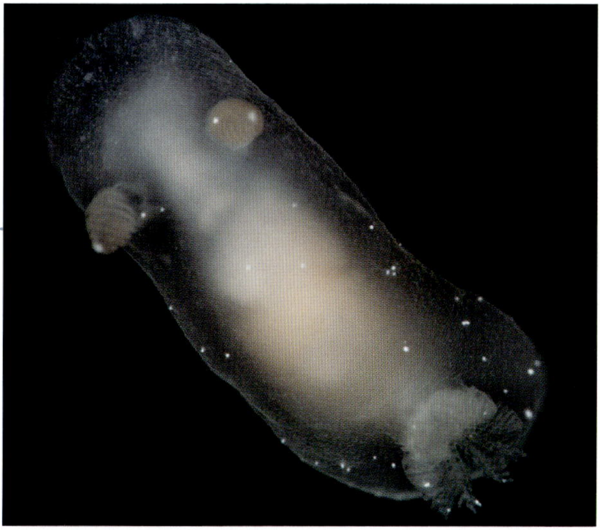

Hallaxa michaeli (Actinocyclidae), Cape Paterson, Vic. Leon Altoff.

Family Aegiridae

A diverse family of slender species varying greatly in colour and body ornamentation. All species have smooth rhinophores, three pre-branchial **lappets** or papillae that tower over the gills, and high rhinophoral sheaths with varying-sized papillae along the margins. All species are feeders on calcareous sponges and can be very cryptic when on their food. Some species have blue or green **ocelli** along each side and over the body.

Aegires exeches

The slender, drab, variably greenish-grey body is beset with long, knobbly spiculose papillae, those in front of the gills and around the rhinophores being highest. Ocelli exhibit a metallic shine here and there between the papillae, and the animal closely resembles a small piece of dull-coloured, spiky sponge.

Size: 12 mm long, usually much smaller.

Habitat and range: intertidal under stones and on algae, at times not uncommon, to 100 m depth. Eastern Australian states from Qld to SA, including Tas; also tropical Pacific.

Family Cadlinidae

Generally firm, rather flattened, small dorids, ovate in shape with tuberculate mantle. Internally: highly denticulate radular teeth, including in *Cadlina* a central tooth with denticulate cutting edge. Cold-water species are muted in colour, mostly whitish with yellowish spots and a yellowish margin on the mantle.

Cadlina tasmanica

A uniformly semitransparent, whitish species, the mantle dotted all over with yellow pigment. The rhinophores and gills are the same as the body colour.

Size: 9 mm.

Habitat and range: intertidal to 6 m depth; under stones. Vic. and Tas.

Aegires exeches (Aegiridae), Aireys Inlet, Vic. Platon Vafiadis.

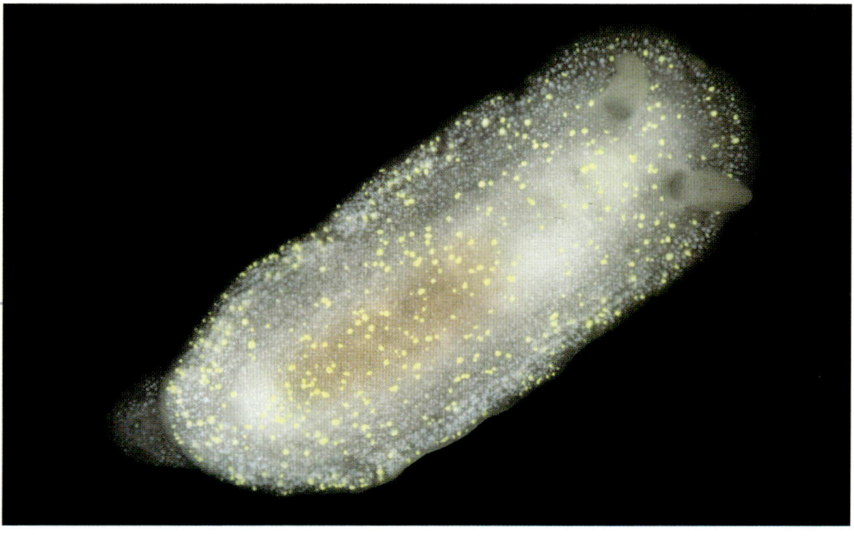

Cadlina tasmanica (Cadlinidae), West Head, Flinders, Vic. Platon Vafiadis.

Family Chromodorididae

A very diverse group of genera with numerous species worldwide, mostly found in tropical and warm-temperate shallow seas. All are sponge feeders, some very exclusively so. Includes some of the most strikingly patterned and coloured nudibranchs, as well as some very drab and uniformly coloured species. Most species are oval in outline, soft, smooth and glossy; a few are elongate and high, quite solid, and ornamented with large papillae and ridges.

Ceratosoma amoenum

A smooth species of distinctive (though somewhat variable) colour pattern. Body pinkish cream, the mantle margin white, interspersed with large or small orange patches, and the median part of the mantle with a few large to numerous small orange spots, among which a scattering of a few deep-red dots can be seen. The rhinophores and gills are deep red with white stripes.

Size: to 60 mm long.

Habitat and range: 13–25 m depth. Southern Qld to southern WA, including Tas.; also NZ.

Ceratosoma brevicaudatum

A bigger and much more solid, elongated and higher species than *C. amoenum*. The tail is more than one-third the total length. Colour basically pale orange, varying from whitish yellow to deep red, always with red and purple spots (each with a lighter halo) on the mantle, sides and tail. A distinctive short, red glandular lobe projects from the mantle edge behind the gills. Victoria's commonest large nudibranch.

Size: to 120 mm, but usually smaller.

Habitat and range: intertidal to 120 m depth. Northern NSW to central WA, including Tas.

Ceratosoma amoenum (Chromodorididae), Roaring Meg West, Wilsons Promontory, Vic. Glenys Greenwood.

Ceratosoma brevicaudatum (Chromodorididae), St Leonards Jetty, Vic. Julian Finn.

Chromodoris alternata

This small nudibranch cannot be confused with any other, due to its unusual colours and pattern. The mantle margin is bluish grey, broken regularly by short, whitish radial bars. The median part of the mantle is an irregular, elongate patch of orange, through the rear half of which the black viscera show. Pale blue flecks may be present on the orange patch. The rhinophores and gills are whitish or fawn.

SIZE: 20 mm.

HABITAT AND RANGE: intertidal to 60 m depth; common; eats the black sponge, *Dictyodendrilla* cf. *caespitosa*. Central Vic. and Tas. to southern WA.

Chromodoris ambigua

One of a number of southern Australian species with orange or red spots on the median part of the mantle, this species is distinguished by the broad white mantle margin containing a single or double series of purple or mauve spots. Rhinophores and gills fawn or buff.

SIZE: 30–40 mm, but usually smaller.

HABITAT AND RANGE: intertidal, to 20 m depth; uncommon; eats the dark purple sponge *Chelonaplysilla violacea*. Central Vic. and Tas. to SA.

Chromodoris tasmaniensis

This white or creamy-white species has a vivid-white mantle margin enclosing a series of small red to orange spots. The median part of the mantle is dotted with many small or fewer large red spots. There are small red spots on the margin of the foot and tail. The margins of the rhinophoral and gill cavities are always lined red.

SIZE: 60 mm, but usually smaller.

HABITAT AND RANGE: intertidal to 60 m depth; eats the pink sponge *Darwinella rosea*. NSW to SA, including Tas.

Chromodoris alternata (Chromodorididae), Portsea Pier, Vic. Glenys Greenwood.

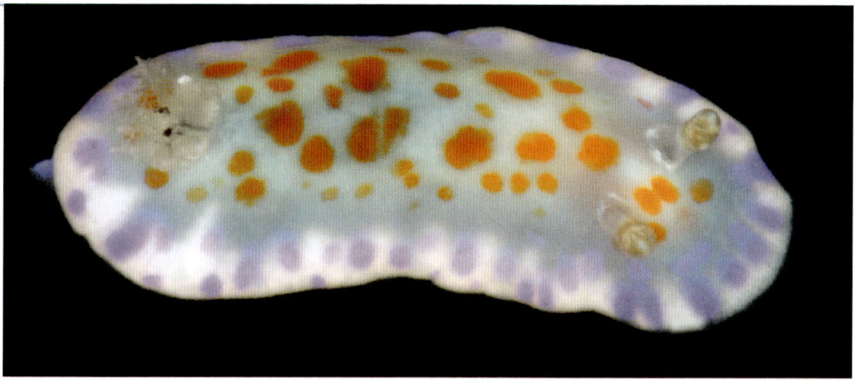

Chromodoris ambigua (Chromodorididae), Point Nepean, Vic. Platon Vafiadis.

Chromodoris tasmaniensis (Chromodorididae), Portsea Pier, Vic. John Chuk.

Diversidoris sulphurea and *Noumea closeorum*

Each of these species has a vivid-yellow body with a submarginal row of orange spots around the mantle. Typically *D. sulphurea* has a row of smaller spots along the edges of the foot and tail, and none on the middle part of the mantle, whereas *N. closeorum* has a few orange spots in the middle part of the mantle. The rhinophores and gills are yellowish in both. There is considerable variation in the pattern in both species, making identification very difficult. The two are separated by internal characters and colour on preservation (pale in *D. sulphurea*, dark brown or black in *N. closeorum*). Both species feed on the same (or similar) sponges.

Size: 20 mm for both species.

Habitat and range: intertidal to 12 m depth. NSW to southern WA (*D. sulphurea*); Vic. and Tas. to SA (*N. closeorum*).

Doriprismatica kulonba

Body, gills and rhinophores semitranslucent white. Mantle and foot narrowly margined with yellow. Opaque white defensive glands visible within the mantle margin. Uncommon.

Size: to 70 mm, usually smaller.

Habitat and range: intertidal to 20 m depth. Vic., Tas. and SA.

Diversidoris sulphurea (Chromodorididae), Blairgowrie Marina, Vic. John Chuk.

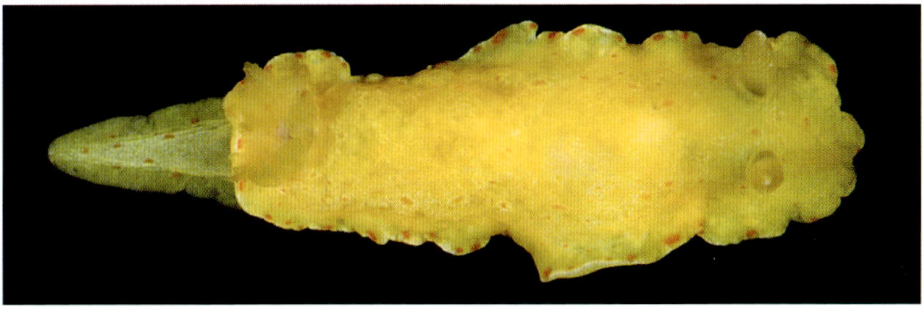

Noumea closeorum (Chromodorididae), Portsea Pier, Vic. John Chuk.

Doriprismatica kulonba (Chromodorididae), Mersey Bluff, Devonport, Tas. Bill Rudman, Australian Museum.

Goniobranchus epicurius

The mantle margin is a narrow, white band enclosing a series of reddish-orange spots. The median part of the mantle is cream or pink, overlaid with a pattern of distinct (Vic. and Tas.) or very crowded (SA) longitudinal white lines, interspersed with a scattering of small orange dots. Rhinophores and gills white to fawn.

Size: 30 mm, often much smaller.

Habitat and range: intertidal to 15 m depth. Vic. and Tas. to southern WA.

Goniobranchus tinctorius

The body is dull white; the mantle has a narrow, deep-yellow margin, inside which is a series of yellow spots. The median part of the mantle is marked by a close, almost honey-comb-like pattern of red lines, with the enclosed areas at midlength reddish. The rhinophores are purplish brown, the gills white with red marks on the ribs.

Size: 40 mm, sometimes much larger.

Habitat and range: rarely intertidal, to 30 m depth. All states of Australia except Tas; also widespread in Indo-Pacific tropical and warm-temperate seas. More than one species may be present under this name.

Hypselodoris bennetti

The body is entirely blue; the mantle is margined with yellow, along the inner edge of which is a series of deep-red spots. Other deep-red spots are present on the median part of the mantle, body sides and foot margins. The rhinophores and gills are bluish with red stripes. The body is high and the mantle overhang narrow.

Size: 40 mm.

Habitat and range: intertidal to 20 m depth. Southern Qld, throughout NSW to central Vic.

Goniobranchus epicurius (Chromodorididae), Flat Rocks, Bunurong Marine Reserve, Vic.
Platon Vafiadis.

Goniobranchus tinctorius (Chromodorididae), Lonsdale Bight, Vic. Andrew Newton.

Hypselodoris bennetti (Chromodorididae), Iron Prince, Vic. ©Jerome Mallefet FNRS-UCL Belgium.

Mexichromis macropus

Radiating orange bars along the mantle margin (with white between) characterise this nudibranch. The body is usually white, but may be mauve. The median part of the mantle has a scattering of high purple pustules, and the rhinophores and gills have a purple tinge. Often seen subtidally in large numbers in southern Port Phillip Bay.

SIZE: 40 mm.

HABITAT AND RANGE: intertidal to 20 m depth; usually on food sponge. Southern Qld to southern WA.

Noumea aureopunctata

A small, dull whitish species, distinguished from small specimens of *Goniobranchus epicurius* by the presence of orange, semicircular spots along the very margin of the mantle.

SIZE: 10–15 mm.

HABITAT AND RANGE: intertidal to 10 m depth. Vic. and Tas.

Noumea closeorum

See text for *Diversidoris sulphurea*.

Noumea haliclona

A wide-ranging, abundant nudibranch exhibiting some variation in colour. The typical colour form is pink with a white margin to the mantle, which is speckled with small red dots. Other colour forms are deep pink with red or orange mantle edge, white with a yellow mantle edge, or even pale yellow with a white mantle edge. Colours are dependent upon the sponge it is eating. Within Port Phillip Bay, the food sponge is consistently the pink *Darwinella rosea*.

SIZE: 15 mm.

HABITAT AND RANGE: intertidal to 20 m depth. Southern Qld to southern WA, including Tas.

Mexichromis macropus (Chromodorididae), San Remo, Vic. Leon Altoff.

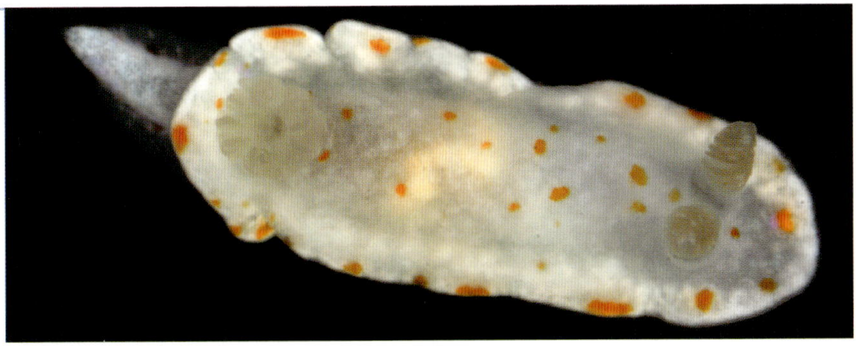

Noumea aureopunctata (Chromodorididae), Point Danger, Torquay, Vic. Platon Vafiadis.

Noumea haliclona (Chromodorididae), Portsea Pier, Vic. John Chuk.

Noumea verconis

The mantle edge is drawn out into an irregular series of pointed lobes that, together with colouration, match exactly the shape and colour of its food sponge, *Darwinella rosea*. Animals are usually vivid-pink or red, rarely an orange pink or bluish yellow. All have a stronger or weaker tracery of pale lines on the median part of the mantle. The tips of the mantle lobes and rhinophores are white.

Size: 65 mm, usually smaller.

Habitat and range: rarely intertidal, to 45 m depth. NSW to southern WA, including Tas.

Thorunna arbuta

An outstanding species, easily recognised by the vivid pinkish-red body colour and three bright-yellow patches along each side of the mantle. The rhinophores and gills are a deeper colour than the body. Body slender with little mantle overhang; anterior foot wider than mantle.

Size: to 30 mm.

Habitat and range: intertidal to 20 m depth. Vic. and Tas. to southern WA.

Thorunna perplexa

A common species, creamy white in colour with orange spots scattered over the median part of the mantle, along the body sides and on the tail. The very edge of the mantle is white, then there is a submarginal row of purple spots or, if **confluent**, a purple line both dorsally and ventrally. The rhinophores and gills have an orange tinge. The body is high and slender, with very little mantle overhang.

Size: 10–20 mm.

Habitat and range: common intertidal to 30 m depth. NSW to Vic. and Tas

Noumea verconis (Chromodorididae) (three individuals side by side), Portsea Pier, Vic. John Chuk.

Thorunna arbuta (Chromodorididae), West Head, Flinders, Vic. Platon Vafiadis.

Thorunna perplexa (Chromodorididae), Point Lonsdale, Vic. Platon Vafiadis.

Family Dendrodorididae

Dendrodorids are soft-bodied nudibranchs, ovate in outline and flatly or moderately convex in profile. They have no hard parts in the alimentary tract. Instead, sponge tissues are dissolved externally by discharged enzymes, and the resulting fluids are sucked in through the pharynx. There are many species worldwide, mostly in shallow water.

Dendrodoris albopurpura

The pale greyish-cream or fawn body has large, purplish-black spots on the median part of the mantle, the surface of which has a few big blister-like pustules. The lamellate club of the rhinophores is bent at an angle to the stem and is purplish brown in colour. The sides of the animal are nearly parallel, the ends broadly rounded. Animals found intertidally appear to have little pedal adhesion, often swirling in the disturbed waters under lifted rocks.

Size: 75 mm.

Habitat and range: under rocks in the intertidal zone to 10 m depth. Central Vic. and Tas. to southern WA.

Dendrodoris arborescens

Always a glossy, velvet-black colour, with a narrow carmine to deep-red marginal band on the mantle edge and the foot. Rhinophores black with white tips; gills black, widely spreading. Often seen crawling across open sand.

Size: 45 mm, but usually smaller.

Habitat and range: intertidal to 10 m depth. All Australian states; very common in Port Phillip Bay; Indo-Pacific.

Dendrodoris aurea

A uniformly pigmented species varying from pinkish to reddish orange, sometimes with a few small, yellow or cream, star-like marks in two series along the mantle. The mantle edges are nearly parallel. Eats a sponge of similar colour.

Size: 60 mm.

Habitat and range: intertidal to 10 m depth. NSW to WA.

Dendrodoris albopurpura (Dendrodorididae), Edithburg Jetty, SA. John Chuk.

Dendrodoris arborescens (Dendrodorididae), Martha Point, Port Phillip Bay, Vic. John Chuk.

Dendrodoris aurea (Dendrodorididae), off Laverton, Port Phillip Bay, Vic. Jan Watson.

Dendrodoris maugeana

The pale-orange body colour, dotted everywhere with small brownish spots, makes this an easily recognised species. The mantle edges are often ruffled. The whole animal has an elongate shape.

Size: 50 mm.

Habitat and range: intertidal to 15 m depth. Central Vic. to southern WA, more common in Tas.

Dendrodoris nigra

Another widespread Indo-Pacific species, one colour form of which occurs at times in Victoria. The tropical form is typically black with a slightly paler mantle margin, and the mantle is peppered or spotted with white dots. Victorian specimens are a paler green-black colour with mantle margin paler again and no white spots.

Size: to 110 mm, usually less than half this length.

Habitat and range: intertidal to 26 m depth. All Australian states except Tas.; Indo-Pacific.

Doriopsilla carneola

This broadly ovate nudibranch is very common. The slightly roughened mantle is very variable in colour, with the typical colour being a dark-red-brown mantle, with the same colour (or a bright-deep-yellow) mantle underside. Common colour forms range from deep red through all shades of orange, pink and yellow to wholly white. Often there are a few small, paler, star-like marks on the mantle. Some colour forms can be confused with *Dendrodoris aurea*, but that species reaches a larger size, is parallel sided, more convex, and softer to touch.

Size: 35–40 mm.

Habitat and range: intertidal to 22 m depth; common in association with a thin, slimy encrusting sponge under shallow-water rocks. Southern Qld to southern WA.

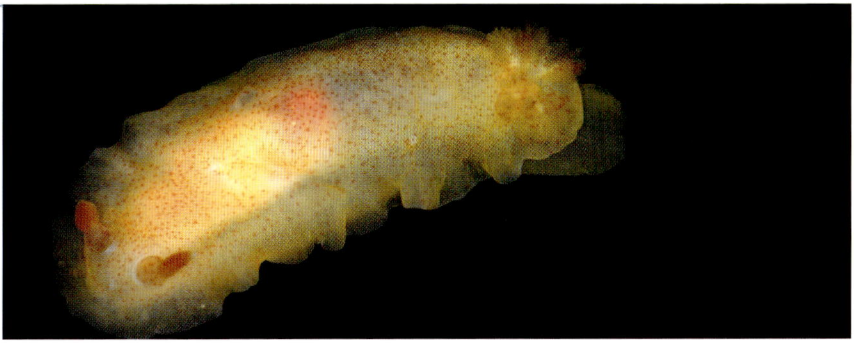

Dendrodoris maugeana (Dendrodorididae), Bridgewater Bay, Vic. Platon Vafiadis.

Dendrodoris nigra (Dendrodorididae) with egg ribbon, Inverloch, Vic. T. Joan Hales.

Doriopsilla carneola (Dendrodorididae), Portsea Pier, Vic. John Chuk.

Doriopsilla peculiaris

A larger, more convex, softer species than *D. carneola*, usually yellowish in colour with opaque, yellow, slightly raised tubercles on the mantle. Some animals are greyish cream with pale-brown patches, others very red.

Size: 40–60 mm.

Habitat and range: intertidal to 30 m depth; more likely to be seen subtidally than intertidally. Central Vic. and Tas. to southern WA.

Family Discodorididae

Discodorids are ovate in shape, lowly convex, with a wide mantle overhang and generally narrow foot. The mantle, which feels rough, is covered with small papillae. In some genera, minute glassy spicules project from the papillae. In the genera *Jorunna* and *Rostanga*, the papillae are cylindrical and close. In *Discodoris* and *Sclerodoris*, they are low, conical and well spaced. The upper lip of the anterior foot is notched below the mouth, and the oral tentacles are laterally grooved.

Alloiodoris marmorata

A large dorid with raised margins to the rhinophoral and gill openings; mantle surface velvet-like. Colour dull fawn with deeper-fawn, often dark-centred, spots and patches.

Size: to 110 mm.

Habitat and range: intertidal to 5 m depth; sand, shingle and algal bottom. Tas. and, questionably, western Vic.

Doriopsilla peculiaris (Dendrodorididae), Portsea Pier, Vic. John Chuk.

Alloiodoris marmorata (Discodorididae), Kingston Boat Ramp, Tas. Sarah Quine.

Discodoris paroa

Always deep bright red in colour, with blackish or brownish specks, spots and mottles patterning the mantle. Rhinophores and gills deeper reddish, their openings large.

Size: to 30 mm. Recently (2010) assigned to the genus *Montereina* but that provisional usage is not followed here.

Habitat and range: intertidal to 48 m depth; usually living and feeding on sponges of the same colour. Vic.

A number of other *Discodoris* species occur in south-eastern Australia, from the intertidal zone to 100 m depth or more, but all are known from very few specimens.

Hoplodoris nodulosa

This is a broadly ovate, rather flat animal with rounded pustules covering the whole mantle. The pustules are largest each side of the midline. The colour varies from pale yellow to dark reddish brown, usually with a lighter reddish-orange patch between the brown rhinophores and the spreading body-coloured gills. Some specimens have cream or yellow pigment between the larger pustules.

Size: to 40 mm, but usually 20–30 mm.

Habitat and range: intertidal under stones, to 55 m depth. Qld to WA, including Tas.; also NZ.

This species name possibly conceals a complex of cryptic species.

Jorunna hartleyi

A uniformly pale-pink species with two or three brownish spots either side of the midline. Mantle covered with slender, cylindrical, close papillae, presenting a carpet of minute spicules. The rhinophores and gills are pale pink, the latter standing up in a circle.

Size: to 40 mm.

Habitat and range: intertidal to 10 m. Vic. and Tas.

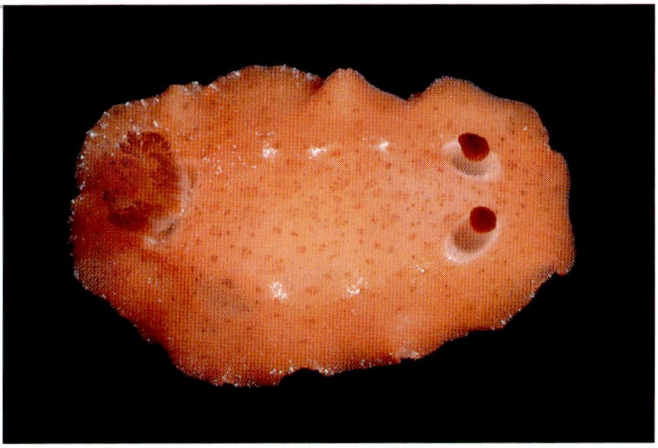

Discodoris paroa (Discodorididae), Portsea Pier, Vic. John Chuk.

Hoplodoris nodulosa (Discodorididae), Flinders Pier, Vic. John Chuk.

Jorunna hartleyi (Discodorididae), Cowrie Beach, Phillip Island, Vic. Platon Vafiadis.

Jorunna ramicola

Similar to the preceding species, but somewhat flatter and smaller. Colour silvery white, mantle patterned with numerous closely spaced unpigmented spots, each surrounded by a ring of white-capped spiculose papillae. The rhinophores and gills are white, the latter not very prominent.

Size: 20 mm or smaller.

Habitat and range: intertidal to 5 m depth. Vic., also NZ, and reported from various parts of the Indo-Pacific.

Another species, *Jorunna pantherina* (not included), also occurs uncommonly in Victorian waters. It has a plumper body than those other species of *Jorunna* figured here, and is uniformly browny grey with some darker patches on the mantle. The rhinophores and gills are fawn, the latter standing up in a high circle.

Size: to 35 mm.

Habitat and range: intertidal to 5 m depth. Southern Australia from Qld to WA; also NZ.

Jorunna sp. RB1

A very distinctive species, easily recognised by its creamy-yellow body, the numerous big black spots on the mantle, the black rhinophores, and the black-outlined white gills. Thus, this species is rather more 'panther-like' than *Jorunna pantherina*! The whole of the mantle is covered with slender, cylindrical, closely spaced papillae, from which project minute, glassy spicules.

Size: to 40 mm, but often smaller.

Habitat and range: lower intertidal to 25 m depth. Southern Qld to southern WA.

Jorunna ramicola (Discodorididae), Blairgowrie, Vic. Glenys Greenwood.

Jorunna sp. RB1 (Discodorididae), Wreck Beach, Vic. Leon Altoff.

Paradoris dubia

A flat, ovate species with a granular mantle surface. Typically a dull blotchy grey animal with a row of two or three small yellowish star-like marks either side of the mantle midline. The rhinophores are dark brown, the gills brownish. The underside of the mantle and foot sole usually has a few brown spots scattered about. Rare individuals have a pure white mantle, with or without the stars.

Size: to 40 mm.

Habitat and range: under stones in the intertidal zone, to 50 m depth. Southern NSW to southern WA, including Tas.; also NZ.

Platydoris galbana

A flattened, leathery species, the mantle minutely granular, foot narrow; body dull yellowish; mantle and foot darker along margins; rhinophores and gills dark reddish brown.

Size: 40–50 mm.

Habitat and range: intertidal to 30 m depth; under stones. Vic. and southern NSW.

Listed as a protected species under the Flora and Fauna Guarantee Act, Vic.

Rostanga calumus

Always a vivid red species; often with marginal and submarginal **concretions** along the mantle edge; often a band of cream-tipped papillae between the rhinophores. Gills reddish; rhinophores cream.

Size: 20–25 mm, often smaller.

Habitat and range: intertidal to 30 m depth; often found under stones, or crawling over algae in pools. Southern Australia from NSW to WA, including Tas.

Paradoris dubia (Discodorididae), Honeysuckle Point, Vic. Leon Altoff.

Platydoris galbana (Discodorididae), Bass Point, NSW. Kirsten Benkendorff.

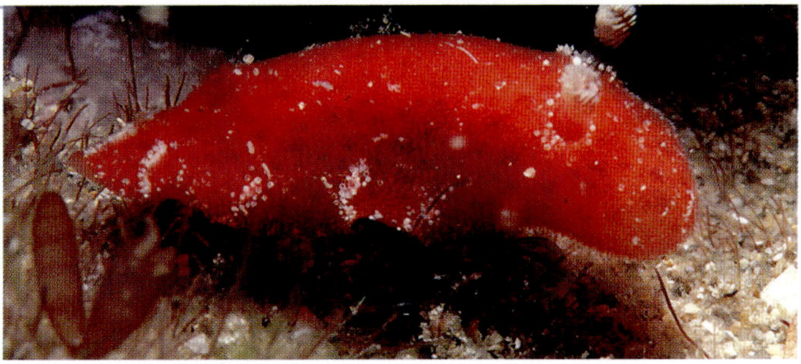

Rostanga calumus (Discodorididae), Portsea Pier, Vic. John Chuk.

Rostanga crawfordi

Colour varies considerably in this species. The typical form is dull orange red, but it may be palest grey, bluish grey or pinkish grey; mantle submargin with a band of white or orange concretions. A low ridge is present in the midline between the rhinophores and gills. The rhinophores are the same colour as the body or darker, short and thick with more or less vertical lamellae near the top. The gills are body coloured and upstanding.

This species appears in earlier publications under several names: it was originally described as *Discodoris crawfordi* and also as *Discodoris turia* and *Rostanga australis*; however, all these records are now known to represent a single species, *Rostanga crawfordi*.

SIZE: 30 mm.

HABITAT AND RANGE: intertidal to 20 m depth; usually associated with its food sponge of the same colour. Vic. and Tas. to WA.

Other species of *Rostanga* occurring in south-eastern Australia are: *Rostanga bifurcata*, with a shallow mid-dorsal crest, plain red in colour, and its spiral egg ribbon very tightly folded; and *Rostanga bassia*, which is dull orange red in colour, its short egg ribbon containing a few large eggs, from which crawling young hatch.

Sclerodoris tarka

This broadly ovate dorid is pale pinkish to brownish grey in colour, with two or three brown spots in shallow pits along the midline between the rhinophores and gills. Low diagonal ribs, made up of slightly higher, slender cylindrical papillae, criss-cross along either side between the midline and the mantle margin. At the intersections of the ribs stand conical papillae, including some larger ones that are capable of much extension. The rhinophores and gills are the same colour as the body.

SIZE: 30 mm.

HABITAT AND RANGE: intertidal to 43 m depth; difficult to see on its food sponge, which it closely resembles. Vic., Tas. and SA.

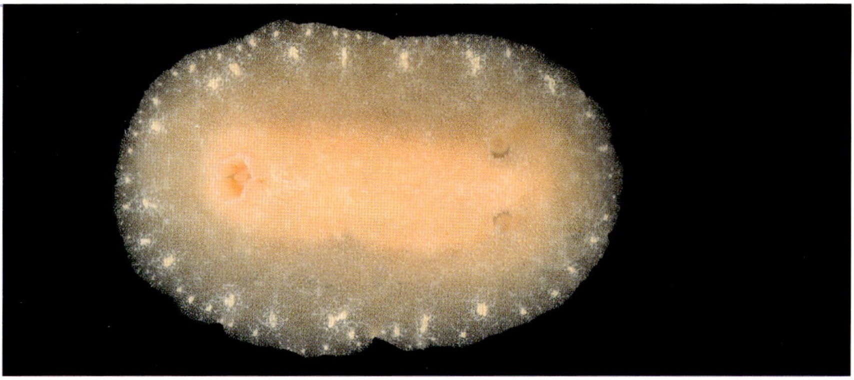

Rostanga crawfordi (Discodorididae), Portsea Pier, Vic. John Chuk.

Rostanga bifurcata (Discodorididae), two individuals with egg ribbon, Wilsons Promontory, Kersops Cove, Vic. Glenys Greenwood.

Sclerodoris tarka (Discodorididae), Blairgowrie Marina, Vic. John Chuk.

Sclerodoris trenberthi

Usually reddish or reddish orange in colour, often with yellowish-cream tips to the irregularly shaped mantle tubercles. Rhinophores and gills dark brownish. Whole animal rather flattened, with thin mantle edge. The mantle tubercles stand in a line either side of the midline.

SIZE: 20 mm.

HABITAT AND RANGE: intertidal, under stones, to 10 m depth. Vic. and SA.

A third species, *Sclerodoris* sp. RB1, has rarely been found in the intertidal zone in Victoria. It is bigger and plumper than the other species, and dull yellowish in colour with a few wispy, brownish marks on the mantle.

Tayuva lilacina

This species is ovate and flattened, with lighter and darker grey blotches on the mantle and long tentacles beside the mouth. The name *Discodoris palma* is a recently recognised synonym.

SIZE: up to 60 mm in length.

HABITAT AND RANGE: intertidal and shallow subtidal; central and eastern Vic., and common in NSW and southern Qld; widespread elsewhere in tropical and warm-temperate waters.

Sclerodoris trenberthi (Discodorididae), Portsea Pier, Vic. Glenys Greenwood.

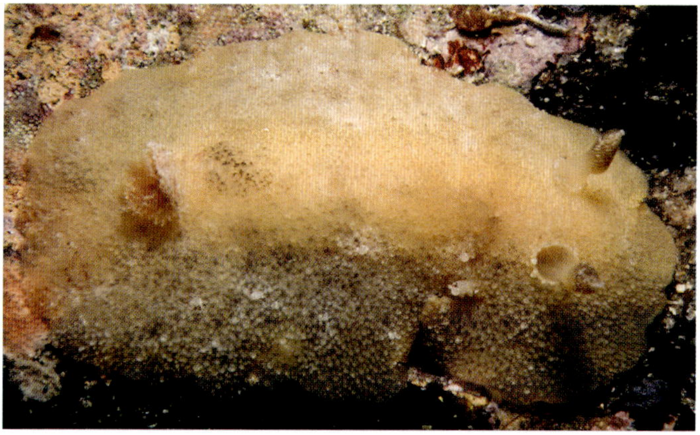

Sclerodoris sp. RB1 (Discodorididae), Portsea Pier, Vic. John Chuk.

Tayuva lilacina (Discodorididae), Blairgowrie Marina, Vic. John Chuk.

Thordisa verrucosa

Generally deep yellow all over, though the mantle may be darker orange to reddish brown. A few large reddish-brown spots mark the underside of the mantle and sides of the body. Elongate conical papillae are scattered over the mantle; these are sticky and usually covered in sand grains. The rhinophores are darker and the gills lighter brown.

Size: 40 mm.

Habitat and range: intertidal zone to 65 m depth; occasionally found under or beside intertidal rocks, partly buried in sand. Qld to Vic.

A second, somewhat smaller, species of *Thordisa* is rarely found in Victorian shallow subtidal waters. This is *Thordisa sanguinea,* with bright red body and the mantle spiked with long, slender, white conical papillae. Another uncommonly seen dorid is *Trippa albata,* a small (15-mm) pure white species, the mantle of which resembles the bristles of a toothbrush.

Family Dorididae

The species grouped in this family are generally ovate in outline, convex in section and smooth surfaced, even when the mantle is formed into nodules or pustules. More difficult to see is the entire, unnotched, upper lip of the anterior foot. There are many species worldwide, including some more than 200 mm in length.

Aphelodoris berghi

A slender, soft species with high collars to the rhinophoral and gill openings; mantle distinctly pustulose (pustules soft and variably sized); colours complex: bluish-grey patches along each side, with maroon or dark brown along the midline and some light-fawn patches marginally. The rhinophores are dark brown, the gills greyish to dark brown. This species can swim by flexing the body.

Size: to 45 mm.

Habitat and range: intertidal to 30 m depth; not common. Vic., Tas. and SA.

Thordisa verrucosa (Discodorididae), Portsea Ocean Beach, Vic. John Chuk.

Aphelodoris berghi (Dorididae), Portsea Pier, Vic. John Chuk.

Aphelodoris rossquicki

Differs from the previous species in broader shape and paler colouration. The mantle edge has a wide cream or fawn band broken by intrusions of the median area pattern of dark-brown wriggly lines and reddish-brown colour. Irregular patches of cream or fawn may break up the median area. The foot is translucent orange with an orange margin. Rhinophoral and gill collars high.

Size: 50 mm or less.

Habitat and range: uncommon, intertidal to 50 m depth. Vic. and Tas.

Aphelodoris varia

Co-marginal brown lines within the mantle edge separate this species from *Aphelodoris berghi* and *Aphelodoris rossquicki*. A pattern of brown anastomosing lines fill the middle part of the mantle, but these may be weak or partly or entirely missing.

Size: to 30 mm.

Habitat and range: intertidal, often in numbers under rocks, to 30 m depth. NSW, eastern Vic. and north-eastern Tas.

Aphelodoris sp. RB1

A large soft species; uniformly orange with brighter narrow mantle and foot margins and purplish-brown rhinophores and gills. A few washy-brown patches may occur on the mantle.

Size: 50 mm or less.

Habitat and range: uncommon, intertidal to 10 m depth. Eastern and central Vic.

Aphelodoris rossquicki (Dorididae), Popes Eye, Port Phillip Bay, Vic. © J. Mallefet FNRS-UCL Belgium.

Aphelodoris varia (Dorididae), Port Stephens, NSW. Heather McLennan.

Aphelodoris sp. RB1 (Dorididae), Beware Reef, Vic. Don Love – Friends of Beware Reef.

Doris cameroni

This nudibranch is broadly ovate and flatly convex. The mantle is covered with small capstan-like pustules, each with a dark spot on the upper surface. Each rhinophoral cavity has a pair of slightly larger protective pustules. Animals vary in colour from pale fawn to dirty greenish brown, often with a darker triangular patch in front of the gills. Rhinophores and gills same as body colour.

Size: 30 mm, often smaller.

Habitat and range: intertidal to 60 m depth. Northern NSW to southern WA.

Several additional species are similar to *Doris cameroni*, but are uncommon or rare. The most similar lacks the black spot in each mantle pustule, the rarest has flattened pustules, each with up to five black spots. Another rare species, *Doris flabellifera*, has flattened gills arranged like a fan towards the tail.

Doris chrysoderma

The whole animal, including rhinophores and gills, is vivid deep yellow, the mantle covered with larger and smaller hemispherical pustules, either white-capped or capped with a yellow shade different to the body colour.

Size: 30–35 mm.

Habitat and range: intertidal to 50 m depth; NSW to WA, including Tas.

Doris wellingtonensis

The largest dorid nudibranch in south-eastern Australia and New Zealand. Ovate in shape and very convex, with large, dome-shaped tubercles all over the mantle, largest in the median area; animal usually yellowish or brownish orange. The large gills are paler than the body colour. The rhinophores are body coloured or darker.

Size: to 200 mm, Australian specimens usually smaller.

Habitat and range: intertidal to 50 m depth. Central and eastern Vic. and Tas; also NZ.

Doris cameroni (Dorididae), Inverloch, Vic. T. Joan Hales.

Doris chrysoderma (Dorididae), Jervis Bay, NSW. Neville Coleman.

Doris wellingtonensis (Dorididae), Devonport, Tas. Bill Rudman, Australian Museum.

Family Goniodorididae

Goniodorids have a very small mantle; the remnants of the mantle edge may form a narrow pallial flange, short or long projecting papillae (along the pallial line vand in the centreline of the mantle), or protective papillae (standing beside the gills and each rhinophore). The gills and rhinophores are contractile, but not retractable within cavities in the mantle. Most Victorian species are small to minute.

Ancula mapae

A small, slender species, creamy yellow in colour, with a prebranchial saddle of rich-brown spotting; gills and rhinophores same colour as the body. From each side of the gills projects a ceras-like, slightly curved papilla; at the outer base of each rhinophore is a deeply bifid papilla.

Size: 4–5 mm.

Habitat and range: 0–10 m depth; crawling on algae in lower intertidal zone. Vic. and NSW. Probably eats nodding heads (kamptozoans). Two other similarly sized species of *Ancula*, both unnamed, occur in Vic. waters: *Ancula* sp. RB1 is pure white, while *Ancula* sp. RB2 has a brown-speckled, dull-orange body and an orange band on each rhinophoral and gill papilla.

Goniodoridella savignyi

Another small species, rather flat dorsally, with the **pallial brim** projecting as a connected series of blunt points culminating each side behind the gills in a pair of longer papillae. Body white; **pallial points** and post-branchial papillae yellow.

Size: 4–7 mm.

Habitat and range: 0–10 m depth; crawling on algae in lower intertidal zone. Eastern Australia from Qld to central Vic.; also widely distributed throughout the Indo-Pacific.

Ancula mapae (Goniodorididae), West Head, Flinders, Vic. Platon Vafiadis.

Goniodoridella savignyi (Goniodorididae), Blairgowrie Marina, Vic. John Chuk.

Goniodoris meracula

The biggest of the local goniodorids. Foot thin-edged and much broader than dorsal surface, the latter having an upstanding, narrow pallial brim. Rhinophores lamellate, stout and cylindrical. Gill large and wide-spreading. Colour variable, from opaque white to deep fawn with brown mottling and spotting.

Size: to 20 mm, but usually smaller.

Habitat and range: 0–10 m depth. NSW, Vic. and SA. Usually found in pits eaten into colonial ascidians (*Botrylloides*). The broad foot lines the pit, and the animal lays its conspicuous egg ribbon along the edge of the pit.

Okenia echinata

A broadly ovate species with many long, conical papillae projecting from the lateral and frontal margins and the dorsal surface between rhinophores and gills; papillae white-tipped above a wide orange band; velar margin and tail orange; body colour greenish brown with white specks.

Size: 8 mm.

Habitat and range: intertidal, under stones. Central Vic. to NSW; western Pacific peripheral temperate waters.

Okenia mija

A small species characterised by the presence of a single midline papilla behind the gills. Along each side of the slender body are six projecting papillae. Body colour pale brownish orange with fine sparse dark-brown speckling all over. Rhinophores with four semicircular lamellae close to the top.

Size: 4–8 mm.

Habitat and range: 0–15 m depth; among algae in lower intertidal rock pools. Central NSW to southern Tas. and central Vic.

Goniodoris meracula (Goniodorididae) with egg ribbons, Portsea Pier, Vic. John Chuk.

Okenia echinata (Goniodorididae), San Remo, Vic. John Chuk.

Okenia mija (Goniodorididae), Blackmans Bay, Derwent River, Tas. Bill Rudman, Australian Museum.

Okenia plana

A very flat, broadly ovate species with a few long, thin, cylindrical dorsal papillae delimiting the pallial line. Pale greyish all over with fine brown speckling.

Size: 8–10 mm.

Habitat and range: 0–3 m; thought to eat the encrusting bryozoan *Membranipora* growing on brown algae and the larger seagrasses. Qld to eastern Vic.; widely distributed throughout the Pacific. A very occasional visitor to the estuaries of eastern Vic.

Okenia rhinorma

This peculiarly shaped species has a very high body tapering to a rather small dorsum, from the margin of which project only six long, conical papillae, plus dorsally the long rhinophores and high, tightly clustered gills; body semitranslucent with dense cream-and-brown stippling and some touches of orange. Lays a drawn-out spiral egg ribbon anchored at one end into the substrate.

Size: 18 mm.

Habitat and range: 5–9 m depth; burrowing through soft sediment in search of its food, the sand ascidian *Mogula*; central Vic. coastline; Indo-West Pacific.

Okenia sp. RB1

The rubbery body of this species is semitranslucent cream with dark-brown spotting on the back and sides, and it has five short stiff papillae, all yellow tipped, along each pallial line.

Size: 4–5 mm.

Habitat and range: 0–13 m depth; associated with the bryozoan *Amathea* growing on brown algal fronds. NSW, Vic. and Tas.

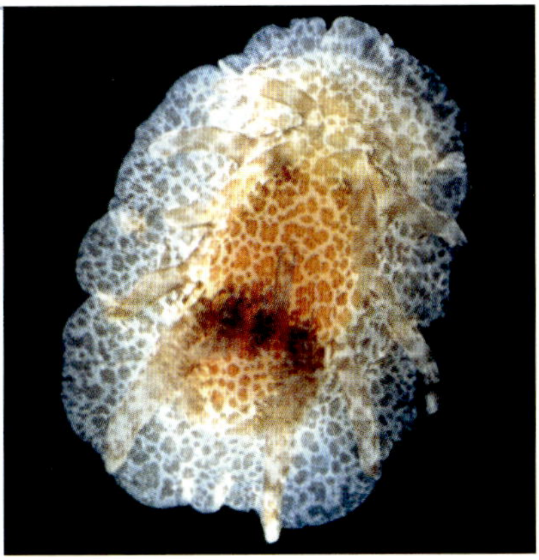

Okenia plana (Goniodorididae), Five Dock Point, Drummoyne, Sydney, NSW.
Bill Rudman, Australian Museum.

Okenia rhinorma (Goniodorididae), Blairgowrie Marina, Vic. John Chuk.

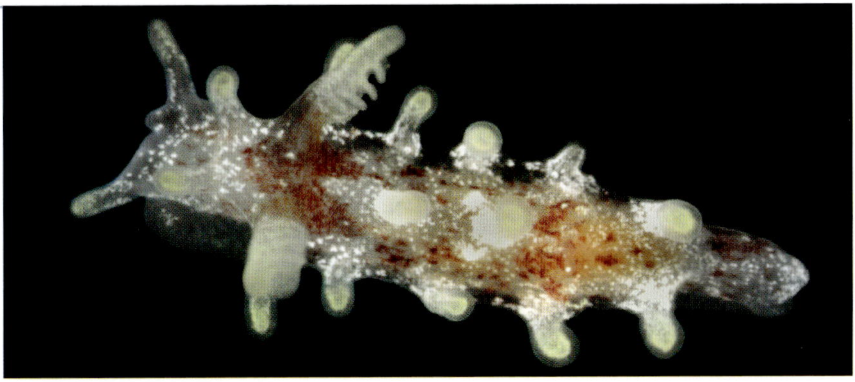

Okenia sp. RB1 (Goniodorididae), Point Lonsdale, Vic. Platon Vafiadis.

Okenia sp. RB2

This very small species is cream with pale-brown spotting and has five short, soft, droopy papillae, white or colourless, along each pallial line.

Size: 4–5 mm.

Habitat and range: lower intertidal and shallow subtidal; associated with the bryozoan *Amathea* growing on brown algal fronds. Vic.

Okenia sp. RB3

The wide, flattened back of this species with 12–13 white- or yellow-tipped conical papillae projecting from the pallial line on each side of the body separate this species from its local congeners. The body is creamy white or fawn, with brown spotting on the back.

Size: 4–8 mm.

Habitat and range: 0–55 m depth; intertidally found crawling on brown algae. Vic. and central NSW.

Genus *Trapania*

This genus is characterised by the presence of a single, usually curved, papilla on each side of the gills, and a single, usually curved, papilla at the outer base of each rhinophore. The body is slender and smooth, usually high just in front of the gills. Animals are known to eat nodding heads (kamptozoans). The three named species known from Victorian waters are 6–16 mm in length and are usually found in shallow subtidal waters.

Trapania aureopunctata

White body with sparsely scattered small golden-yellow spots.

Size: 6–8 mm.

Habitat and range: 2–6 m depth; among nodding heads (kamptozoans) growing on sponges. NSW and Vic.

Okenia sp. RB2 (Goniodorididae), Point Lonsdale, Vic. Platon Vafiadis.

Okenia sp. RB3 (Goniodorididae), Popes Eye, Vic. Platon Vafiadis.

Trapania aureopunctata (Goniodorididae), Portsea Pier, Vic. John Chuk.

Trapania benni

Creamy-yellow body with heavy blotching of light brown, the latter speckled with light blue and yellow.

SIZE: 6–10 mm.

HABITAT AND RANGE: 2–8 m depth; among nodding heads (kamptozoans) growing on sponges. NSW to Vic. and SA.

Trapania brunnea

Usually deep chocolate brown all over with an irregular, cream midline stripe along body and tail.

SIZE: 6–16 mm.

HABITAT AND RANGE: 0–10 m depth; crawling on algae in lower intertidal pools, and among nodding heads (kamptozoans) growing on sponges. Qld to Vic., Tas. and SA; also NZ.

Trapania benni (Goniodorididae), Portsea Pier, Vic. Platon Vafiadis.

Trapania brunnea (Goniodorididae), Portsea Pier, Vic. Platon Vafiadis.

Family Gymnodorididae

Similar in shape to polycerids, but without elongate or branching papillae along the lateral **notal brim**, in front of the head or beside the rhinophores or gills. The gills vary greatly between species in size and disposition: configurations include large and bushy in a circle, small and plate-like with reduced lateral lamellae in a broad arch across the body, and totally absent except for the white branchial glands within the body. Internal characteristics of the alimentary and reproductive tracts also separate the two families. Gymnodorids are carnivores; species of *Gymnodoris* are aggressive predators of other soft-bodied nudibranchs, while species of *Paliolla* eat encrusting bryozoans.

Gymnodoris alba

The slender, white body has a row of small, orange papillae in front of the head and along each side of the body. The yellowish rhinophores are small, bulbous and close together; the yellowish gills are small, flattened and stand in a broad arch or an almost straight line across the body.

Size: 20 mm.

Habitat and range: widespread Indo-Pacific, occurring throughout temperate Australia, uncommon in Vic. Has been observed eating the aeolid *Favorinus* sp. at Blairgowrie, Port Phillip Bay. Victorian specimens may be incorrectly identified, as their transverse row of gills differs greatly from the tight arch of tropical specimens of *Gymnodoris alba*.

Gymnodoris arnoldi

Body pink to reddish brown; rhinophores and gills usually paler than body colour; three gills pointing upwards. The head has a short cylindrical tentacle each side of the mouth and a crest of small, pointed papillae between the tentacles. *Gymnodoris arnoldi* is not to be confused with dark specimens of *Polycera parvula*, which have two papillae on the velar edge in front of the rhinophores. *Rostanga calumus* and *Discodoris paroa* are similar in colour, but have broadly oval body shape with a wide notal overhang all round.

Size: 15 mm.

Habitat and range: intertidal to 60 m depth; Vic. (but uncommon), Tas. and SA. Often associated with the encrusting bryozoan *Beania magellanica*.

Gymnodoris alba (Gymnodorididae), Blairgowrie Marina, Vic. John Chuk.

Gymnodoris arnoldi (Gymnodorididae), Portsea Pier, Vic. John Chuk.

Paliolla cooki

A boxy little species of yellow, orange or pink colour, spotted all over with reddish dots. Rhinophores close together, often appearing darker than in the specimen in the accompanying photograph, or they may be the same colour as the body (as are the gills). The muscular pharynx fills much of the anterior body space and is often visible as a large, red mass within the live animal. Eats the encrusting bryozoan *Beania magellanica*.

Size: 15 mm, but usually smaller.

Habitat and range: intertidal to 130 m depth. Temperate Australia, not common.

Family Onchidorididae

Onchidorids are generally small, ovate, dorid-like nudibranchs, mostly with a mantle so wide that it spreads over the foot to touch the substrate. A few species have higher body sides and a flattened mantle. Some species have a dense network of glassy spicules visible within the mantle. Most species have a dense covering of papillae over the mantle, and some few even have papillae within the gill circlet.

Acanthodoris metulifera

A rarely seen species reported from Victorian intertidal waters. It is larger (to 25 mm in length), brownish green in colour, with sparsely scattered, high, conical papillae all over the mantle. It is more common in southern Tasmania, and may prove identical with the New Zealand species *Acanthodoris mollicella*.

Acanthodoris nanega

A small, flattish, broadly oval species, the mantle of which is covered by distally swollen, crowded, soft papillae. The live animal varies in colour from deep red to rusty reddish brown.

Size: 12 mm.

Habitat and range: 0–3 m depth; under stones in lower intertidal zone. Vic. and SA.

Paliolla cooki (Gymnodorididae), Flinders Pier, Vic. John Chuk.

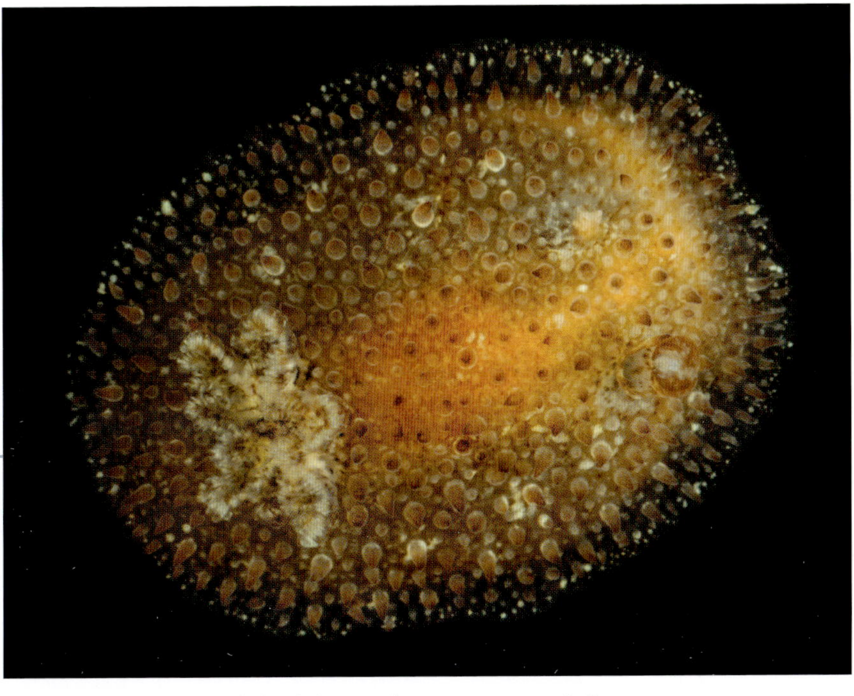

Acanthodoris nanega (Onchidorididae), Eagles Nest, Vic. Leon Altoff.

Corambe sp. RB1

The gills in the species of this genus project into the space between the posterior mantle and the tail. This species has smooth, tapering rhinophores, and the white polygonal pattern on the pale grey mantle exactly matches its host-food, the encrusting bryozoan *Electra pilosa*, upon which it spends its whole life.

Size: 2–4 mm.

Habitat and range: 0–17 m depth; always on its host-food bryozoan encrusting the stems and branches of brown algae. NSW, Vic. and Tas.

Diaphorodoris sp. RB1

A small, elongate-oval species, the mantle sparsely covered with low, broadly conical papillae. Body white or creamy yellow, mantle with a yellow or pale-orange submarginal band.

Size: 12 mm.

Habitat and range: 0–10 m depth; usually crawling over seagrass or brown algae. NSW, Vic. and Tas.

Onchidoris maugeansis

Another small, flattish, broadly ovate, dorid-like species, the mantle covered with high, slender, cylindrical, crowded soft papillae, some of which stand within the gill circle. The rhinophores and gills are almost as high as the papillae. Glassy spicules reinforcing the mantle are conspicuous within the body. Live colouration is pale blotchy yellowish orange, identical to its food.

Size: 8 mm.

Habitat and range: 0–3 m depth; under stones, where it is associated with its food, the encrusting bryozoan *Beania magellanica*. Vic. and Tas.

Corambe sp. RB1 (Onchidorididae), Portsea Ocean Beach, Vic. John Chuk.

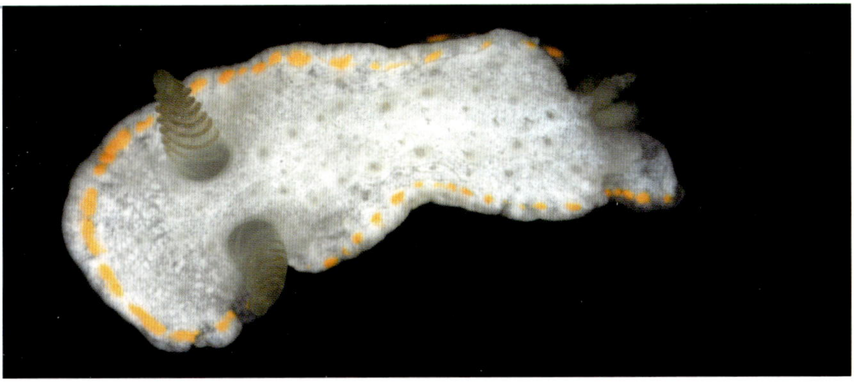

Diaphorodoris sp. RB1 (Onchidorididae), Point Lonsdale, Vic. Platon Vafiadis.

Onchidoris maugeansis (Onchidorididae), Cape Bridgewater, Vic. Platon Vafiadis.

Family Polyceridae

Polycerid nudibranchs are usually long and slender, without a projecting pallial brim. Some species have projecting, smooth or apically digitate processes along either side of the body and in front of the head. The gills are always external. The family is diverse in temperate Australia.

Crimora multidigitalis

The tree-like papillae encircling the body of this species are small and black tipped, the body is pale yellow to orange, and short, conical black papillae are scattered over the back. Often seen in pairs.

Size: 10 mm.

Habitat and range: on algae and bryozoans down to 55 m depth. Central and western Vic. and northern Tas.

Kaloplocamus ramosus

Small, tree-like papillae, usually creamy white in colour, encircle the body of this small, yellow-to-orange species. Species of this genus are known to luminesce when disturbed, and there are many observations of luminescence in this species in Victorian waters.

Size: 15 mm.

Habitat and range: on algae and bryozoans down to 130 m depth. Circumglobal, including Vic.

Crimora multidigitalis (Polyceridae), Portsea Pier, Vic. John Chuk.

Kaloplocamus ramosus (Polyceridae), Portsea Pier, Vic. John Chuk.

Polycera capensis

The largest species of *Polycera* in Australia, introduced from southern Africa many years ago. The white body is marked with wide black stripes along each side and along the midline of the back; six bright-yellow, long, tapering tentacles project along the front edge of the head. A bright-yellow, thicker, longer tentacle stands each side of the black gills. A narrow, bright-yellow line usually marks the division of back from sides, and there is a wider bright-yellow line on the crest of the tail. The rhinophores are black.

Size: 70 mm, but usually smaller.

Habitat and range: 0–10 m depth; usually associated with its bryozoan food. Northern NSW to far eastern Vic.; southern Africa.

Polycera hedgpethi

A larger species with a dark-speckled, grey body; pointed yellow papillae across the head and one each side of the gills. A conspicuous member of the fouling community: originally described from California but now known from many parts of the world, including Australia.

Size: 23 mm long.

Habitat and range: on algae and bryozoans down to 10 m depth. Worldwide including NSW, Vic., Tas., SA and WA.

Polycera capensis (Polyceridae), Beware Reef, Vic. Don Love – Friends of Beware Reef.

Polycera hedgpethi (Polyceridae), Blairgowrie, Vic. © Deb Aston.

Polycera janjukia

This species has bright-yellow spots on a purple to deep-mauve body, a wide head, and large club-shaped rhinophores. The body is smooth, the only projections being six very short, smooth, yellow papillae in front of the head. Often found crawling on algae in pools at low tide.

Size: 15 mm.

Habitat and range: on algae and bryozoans down to 20 m depth. Western Vic. to northern NSW, south-eastern Tas.

Polycera melanosticta

Slender, somewhat translucent body with black speckling all over and small, cream, pointed papillae along the pallial rim each side of the body. Across the front of the head are four longer (usually colourless) papillae, flanked either side by two smaller, shorter, conical papillae. The rhinophores and the tip of the tail are black. Originally described from New Zealand but occasionally common in Victoria.

Size: 20 mm.

Habitat and distribution: on algae and bryozoans to 23 m depth. NSW to central Vic.; NZ.

Polycera parvula

Having just two papillae near the centreline in front of the head and a usually dark-brown or deep-red body distinguish this species. Like *Polycera janjukia*, it is often found crawling on algae in pools at low tide.

Size: 12 mm.

Habitat and range: on algae and bryozoans down to 130 m depth. NSW to central Vic.

Polycera janjukia (Polyceridae), Port Campbell Harbour, Vic. John Chuk.

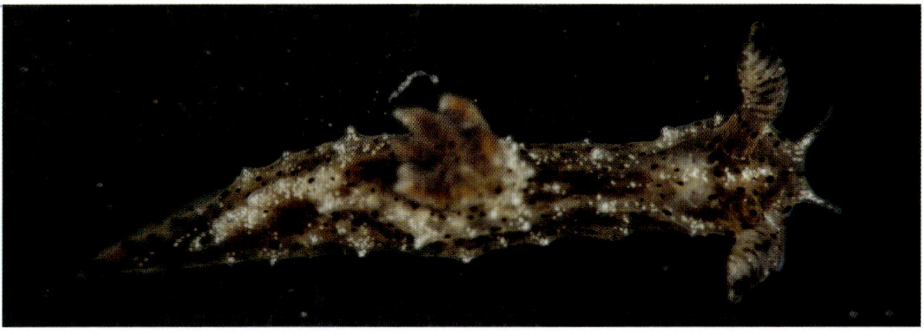

Polycera melanosticta (Polyceridae), Inverloch, Vic. Leon Altoff.

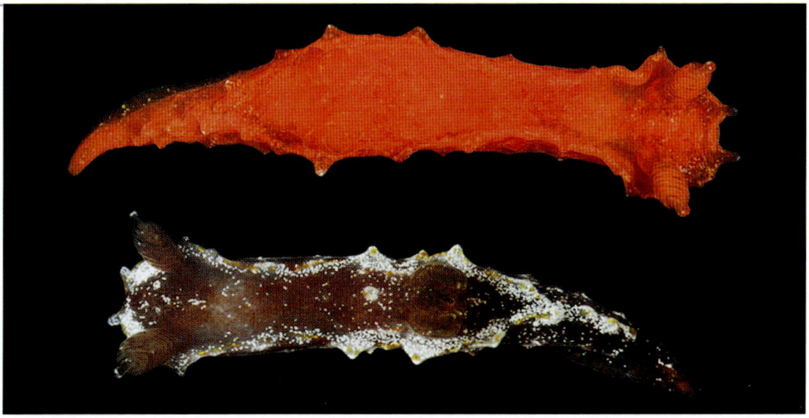

Polycera parvula (Polyceridae): two colour forms, Portsea Pier, Vic. John Chuk.

Polycera sp. RB1

Similar to, but easily separated from, *Polycera parvula* by its usually smaller size, the four short velar tentacles, the semitransparent fawn body colour densely speckled and spotted with cream, orange and black, the small circlet of black gills, and the black tail tip.

SIZE: 5–7 mm, exceptionally to 15 mm.

HABITAT AND RANGE: 0–12 m depth; intertidally found crawling on brown algae in rock pools. Vic. and Tas.

Polycera sp. RB2

This little species is distinguished by the presence of a bulbous appendage each side of the gills. Body pinkish or greyish orange; tips of tail, gill appendages and rhinophores bright orange.

SIZE: 10 mm.

HABITAT AND RANGE: 0–20 m depth; intertidally on brown algae in rock pools. NSW and Vic.

There are at least six other unnamed *Polycera* species known from Victorian waters, including two from deep water.

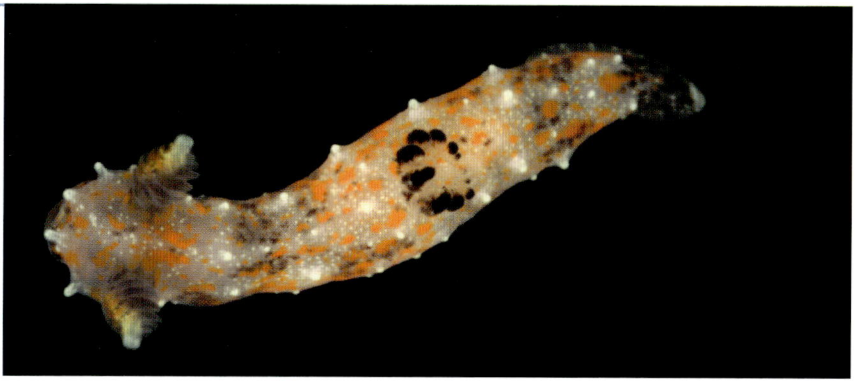

Polycera sp. RB1 (Polyceridae), Point Lonsdale, Vic. Platon Vafiadis.

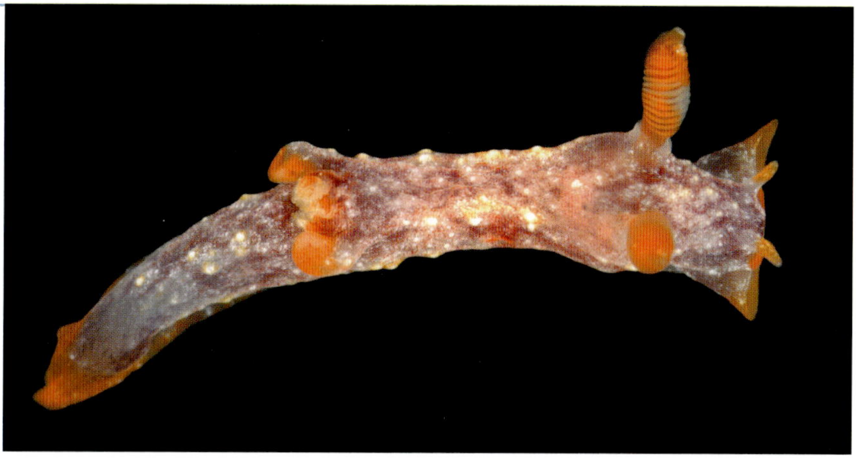

Polycera sp. RB2 (Polyceridae), Point Lonsdale, Vic. Platon Vafiadis.

Tambja verconis

A large, conspicuous nudibranch, well known to divers because of its striking colouration of bright-yellow body with large, deep-blue spots and patches on sides and dorsum, and deep-blue foot margin, pallial rims, rhinophores and gills. Eats the dark-blue-green bryozoan *Bugula dentata.*

Size: 120 mm.

Habitat and distribution: on its host-food to 50 m depth. NSW to WA, including Tas.; NZ.

Tambja sp. RB1

Equally well known to divers, this species has a uniform olive-green to dark-brownish-green body, brown rhinophores, and brown gills with white or cream ribs.

Size: 50–70 mm.

Habitat and range: 0–20 m depth; usually on or close to its food bryozoan, *Bugula dentata*. NSW to SA, including Tas.

Tambja verconis (Polyceridae), Portsea Pier, Vic. Mark Norman.

Tambja sp. RB1 (Polyceridae), Portsea Pier, Vic. John Chuk.

Tambja sp. RB2

The bright-orange body, with purple tips to tail, gills and rhinophores, easily separates this species from its local congeners. May be a warmer-water visitor to our shores.

SIZE: 30 mm.

HABITAT AND RANGE: 0–5 m depth; crawling on brown algae. Vic.

Thecacera pennigera

Generally smaller than members of the genus *Tambja*, this species is white with spotting of orange, yellow and black all over. A large, curled, leaf-like sheath, open behind, protects each rhinophore, and there is a smooth papilla each side of the gills. Another conspicuous member of the fouling community, widely distributed with records from many parts of the world.

SIZE: 25 mm.

HABITAT AND RANGE: on algae and bryozoans down to 10 m depth. Worldwide, including NSW to SA.

Family Vayssiereidae

A family, probably of only the one species, described and named several times from widespread localities. The body is very small and slender with smooth rhinophores, and there are no gills. This nudibranch is always found in association with **serpulid polychaetes**, the calcareous tubes of which it bores into to access the retracted worm, which it then proceeds to eat.

Vayssierea caledonica

Varying in colour from pale orange to pale red, always with darker-red ovotestes showing through the middle part of the body; body and rhinophores completely smooth. Along the Victorian coast found in association with serpulid polychaetes, especially *Salmacina australis*.

SIZE: 3 mm.

HABITAT AND RANGE: intertidal to 3 m depth. All eastern Australian states except Tas.; also tropical and temperate Pacific.

Tambja sp. RB2 (Polyceridae), Aireys Inlet, Vic. Platon Vafiadis.

Thecacera pennigera (Polyceridae), Blairgowrie Marina, Vic. John Chuk.

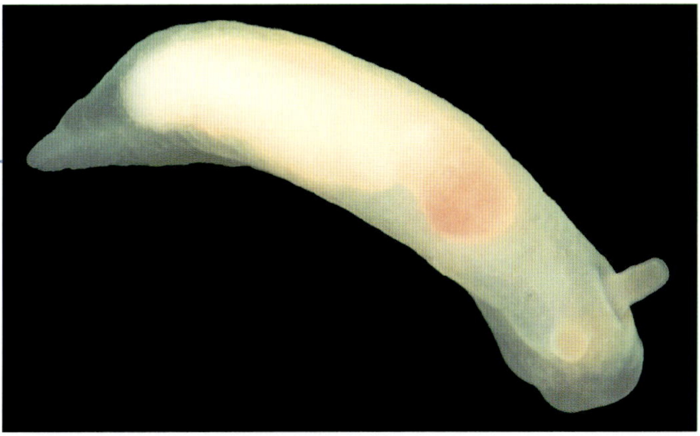

Vayssierea caledonica (Vayssiereidae), Cape Bridgewater, Vic. Leon Altoff.

Suborder Dendronotina

Families grouped within this suborder all have an anterior velum across the head, and high columnar rhinophoral sheaths, each with a smallish distal section (termed a rhinophoral **clavus**) cupped in the top. The rhinophoral clavus is usually short and lamellated, though in Dotidae it is longer and smooth. The body is usually long and slender, often as high as wide. Most species have a series of dendritic processes along each side of the body, but in *Melibe* and Dotidae the processes are **ceratiform** (as in aeolid nudibranchs) and in Scyllaeidae almost parapodia-like (as in the sacoglossan *Elysia*). The majority of species have a radula (absent in *Melibe*), many have strong jaws (including some *Melibe*), and some have a band of hard grinding plates in the stomach. Many species have an obligate association with their soft coral food, upon which they are well disguised by colour and shape. Others live and feed upon hydroids, while *Melibe* casts its velum to catch minute crustaceans. Most species crawl elegantly hither and yon upon their food or in search of it. A few species can 'swim' by violently threshing the body head to tail, lifting off the substrate and being swept along by currents.

Family Dendronotidae

These nudibranchs are very similar in external form to tritoniids, but have densely ramose branchial processes. Otherwise, internal differences separate the two families. Dendronotids eat hydroids and **cerianthid anemones**.

Two very rare species of *Dendronotus* are known from the Bass Strait area. Both are small; one species is pinkish, the other greyish white with yellow spots along the dorsum. Nothing is known of their ecology, except that both species were found in areas of high hydroid diversity. The pinkish species, *Dendronotus* sp. RB1, has been found at 10–80 m depth in Tasmania and Victoria, the white species, *Dendronotus* sp. RB2, at 3–6 m depth in Victoria only.

Dendronotus sp. RB2 (Dendronotidae), Popes Eye, Port Phillip Bay, Vic. Trevor McMurrich.

Family Dotidae

Most species are small to very small and slender, with a rounded velum across the front of the head, smooth rhinophores that each arise within a high, cylindrical sheath with expanded rim, and 5–11 swollen, knobbly cerata along each side of the back. All dotids eat (and lay their winding, wavy egg ribbons on) hydroids. There are many species worldwide, including several colourful undescribed species in south-eastern Australian waters.

Doto ostenta

The small white or cream body may have varying numbers of black patches. The rhinophores are white. The cerata are swollen and knobbly, with two or three circlets of short, rounded projections, the tip of each having a black spot. The digestive duct within the cerata varies in colour, and can be yellowish pink, brownish or deep reddish, depending upon the hydroid being eaten.

Size: 10 mm.

Habitat and range: intertidal to 85 m depth; among hydroids under rocks, or crawling about on algal tips in lower intertidal rock pools; at times very common. NSW to SA, including Tas.

Doto pita

Similar to *D. ostenta*; body ranging in colour from cream overlaid with sketchy, reddish-brown streaks and patches to completely reddish brown; the rhinophores are brown with white tips; the cerata are irregularly craggy in shape and reddish brown in colour over cream digestive glands.

Size: 12 mm, but usually 5–8 mm.

Habitat and range: intertidal to 95 m depth; among hydroids under rocks, or crawling on algal tips in lower intertidal rock pools; subtidally on its hydroid food; common. Vic. and Tas.; also NZ, Japan, Brazil and the Caribbean.

Doto ostenta (Dotidae), Point Lonsdale, Vic. Platon Vafiadis.

Doto pita (Dotidae), Blairgowrie Marina, Vic. John Chuk.

Among the unnamed species of *Doto* known from south-eastern Australia are two very similar to *Doto ostenta* and two larger species with striking colouration. *Doto* sp. RB1 has a white body and cream cerata without a black dot in the tip of each **secondary papilla**. Habitat and range: intertidal to 18 m depth. Vic. and Tas. *Doto* sp. RB4 also has a white body, but the orange cerata are speckled with black. Habitat and range: 30 m depth. Vic. *Doto* sp. RB2 has a white body with a red stripe along the middle of the back, branching to each rhinophore, cerata brownish. Habitat and range: 5–33 m depth, on the hydroid *Gymnangium*. NSW, Vic. and Tas. *Doto* sp. RB3 is the only one of these species for which a photograph was available for this publication; it has large pale-pink or orange spots on the body sides and back with fawn cerata. Habitat and range: 10–40 m depth, on the hydroid *Gymnangium*. Vic. and Tas.

Family Scyllaeidae

Scyllaeids are largish nudibranchs with one or two big, wing-like flaps each side of the body. From the inner face of these flaps project fine, fluffy contractile gills that can fill the space between the flaps. Each rhinophore has a small, lamellate club arising from within the cup-like top of the long cylindrical rhinophoral sheath. Scyllaeids feed upon hydroids growing on algae. There is one cosmopolitan species and several widely spread Indo-Pacific species in the family.

Crosslandia viridis

Easily identified by the single, large, ragged parapodial flap each side of the body, and the (usually uniform) lettuce to dark-green body colouration.

Size: to 20 mm.

Habitat and range: intertidal to 10 m depth; among algae and seagrass. Qld; rarely in Vic.; tropical and temperate Indo-Pacific.

Doto sp. RB3 (Dotidae), Bass Strait, south of Mount Colite (Barwon Heads Bluff), Vic. Platon Vafiadis.

Crosslandia viridis (Scyllaeidae), Dar es Salaam, Tanzania. Bill Rudman, Australian Museum.

Scyllaea pelagica

This green, yellow or brown nudibranch is slender with a pair of elongate flaps with papillae on their margin along each side of the body, high rhinophoral sheaths on the head, and usually small, bright-blue ocelli scattered here and there over the back and sides.

Size: 50 mm.

Habitat and range: intertidal to 10 m depth; usually among seagrasses and floating masses of algae. Cosmopolitan. It has been suggested (2012) that the name *Scyllaea pelagica* be restricted to specimens from the Atlantic Ocean, and that the name *Scyllaea fulva* be applied to those from the Indo-Pacific.

Family Tethydidae

Unique to members of this family is the expanded cowl forming the head. The cowl is used to catch and retain food prior to ingestion. The animal casts the cowl forward, rather like a fisherman casts a throw-net. Each rhinophore has a small, lamellate club arising from the cup-like top of the long, cylindrical sheath; a thin, vertical flange may be present on the upper rear edge of the sheath. Spindle-shaped or wedge-shaped cerata project from each side of the body. Some tropical species exceed 300 mm in length, but most are much smaller. Tethydids lack radular teeth, some have jaws, and most have a band of hard crushing plates in the stomach. When handled, tethydids exude a sweet, sickly scent.

Scyllaea pelagica (Scyllaeidae), West Channel, Port Phillip Bay, Vic. Mark Norman.

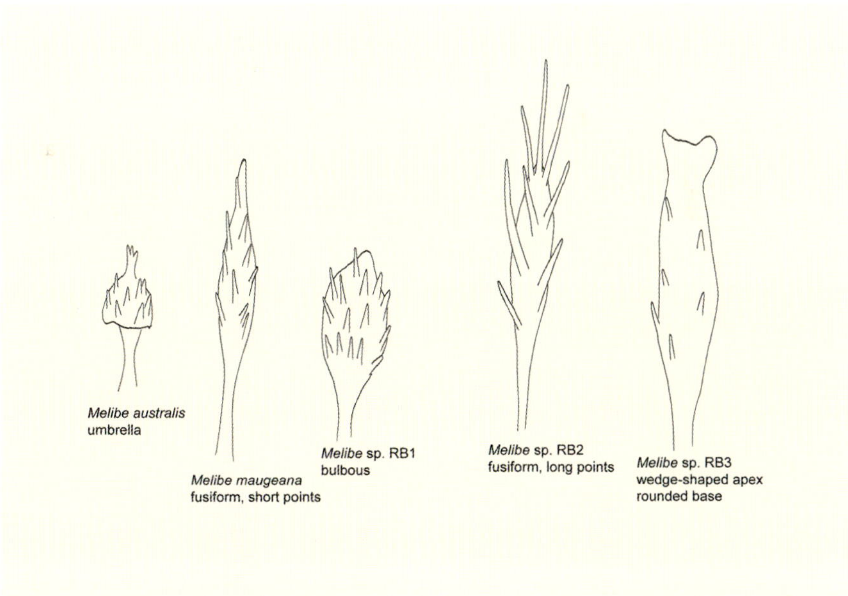

Species of *Melibe* occurring in the Bass Strait area can be readily identified by the shape of the cerata along each side of the body. Drawing: R. Burn, 2014.

Melibe australis

A small, cream species sometimes speckled with red or brown, with short (often mushroom-shaped) cerata, each with a knobbly top and a central apical cylindrical papilla. The rhinophores have short stems; stomach without crushing plates.

SIZE: 25 mm.

HABITAT AND RANGE: intertidal to 10 m depth; usually found under stones or crawling on algae in low intertidal rock pools. NSW to southern WA, including Tas. Possibly more than one species is hidden under this name.

Melibe maugeana

A translucent, yellowish species with a slender, high body and long, spindle-shaped cerata terminating in a single or double point; each rhinophoral sheath has a wing-like flange behind; stomach with a band of strong crushing plates; can swim by vigorous lateral flexing of head to tail.

SIZE: 35–40 mm.

HABITAT AND RANGE: intertidal to 10 m depth; under stones or crawling on algae. Presently known only from Vic.

Three unnamed species of *Melibe* occur in south-eastern Australian waters. *Melibe* sp. RB1 has short, swollen, semitransparent club-shaped cerata covered with finely pointed short papillae; intertidal, Vic. *Melibe* sp. RB2 is like *Melibe maugeana* but has long slender papillae projecting from the cerata and from the rear of the rhinophoral sheaths; intertidal, Vic. and SA. *Melibe* sp. RB3 is greenish orange or brown in colour, the body is slender and angular, and the cerata apically wedge-shaped; 0–10 m depth, Vic.

Melibe australis (Tethydidae), Portsea Pier, Vic. John Chuk.

Melibe maugeana (Tethydidae), Point Leo, Vic. Leon Altoff.

Melibe sp. RB1 (Tethydidae), Point Danger, Torquay, Vic. Platon Vafiadis.

Family Tritoniidae

Worldwide, tritoniids range in size from less than 5 mm to almost 300 mm. All tritoniids have a series of tree-like branched branchial processes along either side of the back, and palmate, sheathed rhinophores. The back may be narrow and the processes relatively few in number, or the back may be broad and dorid-like, with numerous small processes along the edge. The rhinophores and their sheaths closely resemble the branchial processes in colour, shape and size. Internally, the stomach is mostly armed with a band of gastral plates. Tritoniids eat soft corals and are often food specific.

Marioniopsis platyctenea

A small species, amazingly flattened and dorid-like in shape, with numerous very small, branched branchial processes along the very margin of the mantle. The rhinophores and sheaths are small and simple. Body colour is dirty greyish white, often with a mid-dorsal rounded patch of lighter or darker colour, mimicking a soft coral polyp. The rhinophores are brownish.

SIZE: to 20 mm.

HABITAT AND RANGE: shallow subtidal to 20 m depth; always on the encrusting soft coral *Erythropodium hicksoni*, upon which it feeds (leaving visible feeding tracks) and breeds. Temperate Australia from southern Qld to southern WA, including Tas.

Marianina rosea

A small to very small, slender species with a series of double, blade-like cerata along each side. The rhinophore sheath has a single short process projecting from the rear edge. A pair of long, oral tentacles project forward on the head. Body colour bright reddish pink, tending to purplish; oral branchial processes and tail white.

SIZE: 5–10 mm, more often smaller than large.

HABITAT AND RANGE: intertidal to 25 m depth; tropical and temperate Indo-Pacific, south to NSW and SA, western Vic.

At least 10 other tritoniid species are known to occur in south-eastern Australia from the intertidal zone down to 600 m depth. Most are known from only one or two specimens. Mention of four of these species is given below to give some idea of the variation in body shape and colour.

Marioniopsis platyctenea (Tritoniidae) with egg ribbon, Flinders Pier, Vic. John Chuk.

Marianina rosea (Tritoniidae), Noumea, New Caledonia. Bill Rudman, Australian Museum.

Tritonia sp. RB1

Body pale bluish mauve or pink; branchial processes: three each side, brown and strongly branched; velar tentacles: six or seven, long and slender.

SIZE: up to 14 mm, but generally smaller.

HABITAT AND RANGE: on algae in low intertidal pools, to 10 m depth; Vic.

Tritonia sp. RB2

Body fawn to pale brown, often with a series of dark-brown spots along the middle of the body; dorsum with scattered, low, dome-like papillae; branchial processes: four each side, tips ill-formed and twig-like; velum with projecting corners only.

SIZE: up to 12 mm, but generally much smaller.

HABITAT AND RANGE: on algae in low intertidal pools, to 13 m depth; Vic.

Tritonia sp. RB3

Body orange to dark brown with white transverse bands between and up the branchial processes and along the middle of the body; branchial processes: 8–10 each side, tall and branched like a small tree; velar tentacles: 8–10, slender and pointed.

SIZE: up to 25 mm, but mostly smaller.

HABITAT AND RANGE: on algae in low intertidal pools, to 30 m depth; temperate Australia from Qld to WA, including Tas.

Tritonia sp. RB1 (Tritoniidae), Saxon Reef, Portland, Vic. Leon Altoff.

Tritonia sp. RB2 (Tritoniidae), Blairgowrie Marina, Vic. John Chuk.

Tritonia sp. RB3 (Tritoniidae), Portsea Pier, Vic. John Chuk.

Tritonia sp. RB4

Body pale, opaque, white with numerous clear, ovate patches in three series along the back; branchial processes: eight each side, not greatly branching; velar tentacles: reduced in number.

Size: up to 25 mm.

Habitat and range: subtidal, 3–10 m; Vic.

Suborder Arminina

This suborder is a very mixed group. It includes some very uncommon, mostly deep-water species, as well as species found only in association with their host-food. Most species have an anterior velum and a pair of contractile, non-retractile rhinophores without sheaths. Gills, if present, line each side of the body, either laterally in the space between foot and mantle, or dorsally along the edge of the dorsum. The anus opens well towards the posterior end of the body, either on the right side between mantle and foot, or mid-dorsally a little in front of the tail. The body may have an elongate tapering mantle, either smooth or sculptured with ridges or pustules, or there may be lateral tree-like processes along each side, or aeolid-like cerata fringing the dorsum, including across the anterior body.

Armina sp. RB1

The shield-like mantle of this species is dull white, sometimes much darker, with numerous slender, shining, white ridges running from head to tail. Lateral gills present anteriorly only. Broad blackish bands cross the body at about one-quarter and at half animal length and near tail tip. Rhinophores are bright red; foot pinkish. Feeds on the sea pen, *Sarcoptilus grandis*.

Size: 30–40 mm.

Habitat and range: 2–55 m depth; either feeding on or in sediment close to its host-food. NSW to southern WA, including Tas.

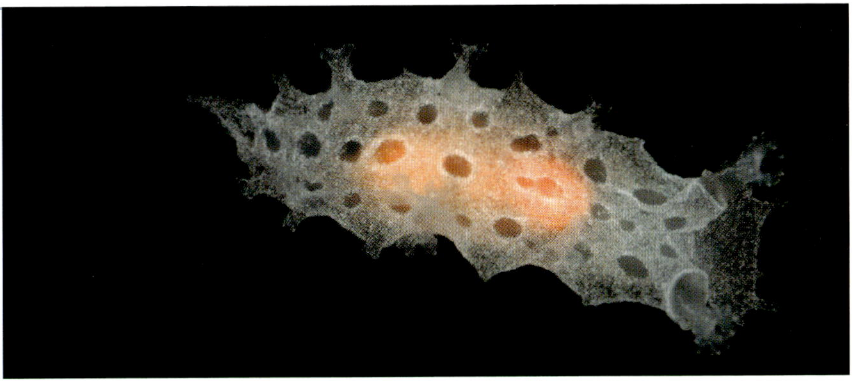

Tritonia sp. RB4 (Tritoniidae), Popes Eye, Port Phillip Bay, Vic. Platon Vafiadis.

Family Arminidae

Species of this family are those with an elongate, tapering mantle and, in some species, lateral gills in the space between foot and mantle. The rhinophores are usually close together, situated in a notch between the anterior ends of the mantle margins. The anus is right posterior–lateral. Arminids feed on sea pens (Pennatulacea).

Armina sp. RB1 (Arminidae) at base of sea pen, Jervis Bay, NSW. Neville Coleman.

Dermatobranchus rubidus

Similar to *Armina* sp. RB1 in shape, but the animal is pinkish and the dorsal ridges are white, the mantle is broader anteriorly with the side margins a little wavy, and the velum is broad with pointed corners.

Size: 20–30 mm.

Habitat and range: 0–30 m depth; in sandy sediment. Vic. and Tas.; northern NZ and wider tropical Pacific. Previously known as *Dermatobranchus pulcherrimus.*

Family Madrellidae

The processes encircling the dorsum of madrellids are elongate and conical. Within each process is an easily discharged sac of viscous pigment, taken up from its bryozoan food and stored for defensive purposes. The rhinophores resemble an upside-down stove brush. The anus opens on the posterior right side.

Madrella sanguinea

An anteriorly broadly ovate animal tapering to a blunt point behind, with numerous elongate conical cerata in a broad band around the body, lateral cerata almost touching the substrate; dorsum between cerata densely pustulose, pustules low. A rusty-red nudibranch with white or cream on (and bridging between) the rhinophores. Eggs hatch as swimming veligers.

Size: up to 33 mm.

Habitat and range: intertidal to 3 m depth; on mixed bottom growths. Central NSW to central Vic. Newly reported from Blairgowrie Marina, Port Phillip Bay, where it lays its long, winding egg strings on the erect bryozoan *Orthoscuticella ventricosa*. Taxonomic revision may show this to be synonymous with *Madrella ferruginosa*.

Dermatobranchus rubidus (Arminidae), Whangarei Harbour, NZ. Richard Willan.

Madrella sanguinea (Madrellidae), Blairgowrie Marina, Vic. John Chuk.

Madrella sp. RB1

Compared with the above species, this is a smaller, more elongate, darker-blood-red animal with far fewer cerata in a narrow band around the body. The body sides are higher, the cerata do not project laterally, the dorsum between cerata is sparsely pustulose, and the pustules are high; upper part of rhinophores orangish. Eggs hatch as crawling young. Handling specimens with bare hands results in blood-like staining to the fingers. This species has long been misidentified as *Madrella sanguinea*, a warm-temperate species with swimming young.

Size: exceptionally 15–17 mm, usually 5–10 mm.

Habitat and range: intertidal to 23 m depth; on its obligate food bryozoan, the dark-red encrusting *Mucropetraliella ellerii*, or crawling on algal tips in rock pools; common. Temperate Australia from central NSW to southern WA, including Tas. Deeper water specimens tend to be paler in colour.

Family Proctonotidae

Proctonotid species have a flattish body encircled by a dense band of smooth or prickly fusiform cerata. The rhinophores have long stalks and lamellate or knobbly upper parts. There may be an inter-rhinophoral ridged crest, raised in some species, flattened in others. The anus opens through a cylindrical papilla towards the rear of the dorsum on the midline. All species feed on bryozoans. Previously known as Zephyrinidae.

Caldukia affinis

An orange and brown-spotted species with no crest between the rhinophores. The cerata are smooth, short and stoutly spindle-shaped with blunt tips; eats a pale orange encrusting bryozoan.

Size: 15 mm.

Habitat and range: intertidal to 55 m depth; often found on its food bryozoan, or crawling on brown algae in deep rock pools. NSW to SA, including Tas.

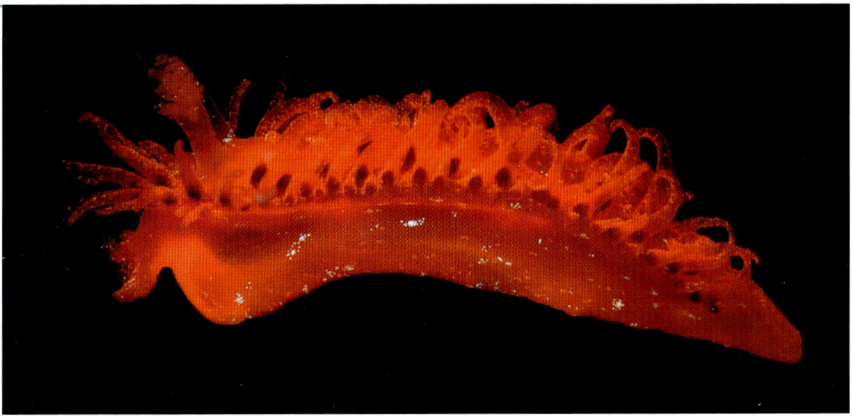

Madrella sp. RB1 (Madrellidae), Portsea Ocean Beach, Vic. John Chuk.

Caldukia affinis (Proctonotidae), Flinders, Mushroom Reef, Vic. Leon Altoff.

Janolus hyalinus

Prominent inter-rhinophoral crest present; cerata elongate pear shaped, each with short projecting points on the upper half. The body colour is brownish with reddish patches, the cerata and rhinophores blotchy creamy brown; eats pale-orange to pale-brown arborescent bryozoans.

SIZE: 30 mm.

HABITAT AND RANGE: intertidal to 18 m depth; found in association with or close to its food bryozoan. Vic.; also NZ and Europe.

Janolus sp. RB1

A large nudibranch with crowded, smooth cerata hiding almost all the body except for the projecting rhinophores; prominent inter-rhinophoral crest present. Body colour deep yellow or orange; foot margins with splashes of cream, and cerata the same colour as the body, each with wide cream band below tip.

SIZE: to 40 mm.

HABITAT AND RANGE: 0–20 m. NSW to southern WA, including Tas.

Janolus sp. RB2

A smaller, slender species with a creamy body and long tail; inter-rhinophoral crest low but distinct; cerata smooth, not crowded, leaving much of the dorsum visible; body with a few brown patches along centreline; cerata with silvery-white lines from tip to base, an orange band below tip; digestive gland brown.

SIZE: to 20 mm, usually smaller.

HABITAT AND RANGE: 0–10 m. Presently known only from central Vic.

Two other orange species of *Janolus* are known from single specimens found along the central Victorian coastline. *Janolus* sp. RB5 has crowded, purple-tipped cerata, while *Janolus* sp. RB3 is more translucent and has orange specks all over body and cerata.

Janolus hyalinus (Proctonotidae), Portsea Ocean Beach, Vic. John Chuk.

Janolus sp. RB1 (Proctonotidae), Point Nepean, Vic. Platon Vafiadis.

Janolus sp. RB2 (Proctonotidae), Point Danger, Torquay, Vic. Platon Vafiadis.

Suborder Aeolidina

Commonly called aeolids, these nudibranchs have cerata (usually in groups) along each side of the body, each ceras containing a duct of the digestive gland and usually having a cavity (cnidosac) to store ingested nematocysts in the tip, which distinguishes aeolids from all other nudibranchs. Some specialised species do not store nematocysts, and a few cultivate zooxanthellae as an energy source in the digestive ducts spreading throughout the body and cerata. Most species feed upon cnidarians, and a few eat nudibranch or fish eggs. Aeolids lack sheaths to the rhinophores. The head has a pair of oral tentacles, which easily separates them from similar-looking cerata-bearing sacoglossans, which only have rhinophores.

Innumerable aeolid species occur in south-eastern Australian waters, the vast majority small; many await a name and description. What little is known of their distribution suggests that many have very restricted ranges. A number appear to be very rare, with 20–30 years between sightings. Almost nothing is known of the aeolid fauna of waters deeper than SCUBA-diver depth.

Aeolidiella drusilla

A large, plump aeolid with pale-pink or greyish-white body and a mat of fawn or buff cerata covering the back. The rhinophores are short and smooth, the anterior foot very wide with large tentacle-shaped corners and the cerata slightly flattened.

Size: 60–70 mm.

Habitat and range: very rare intertidally, to 58 m depth. Vic. and Tas.; also NZ.

Family Aeolidiidae

Generally large and broad aeolids with a close covering of cerata, though one species, *Burnaia helicochorda*, is slender with rather few, well-spaced cerata. The cerata are cylindrical or somewhat flattened, with the first rows beside the rhinophores. The foot is broad, the anterior corners tentacle-shaped. The rhinophores may be smooth, bearing low oblique ridges, densely papillate all round (or on the rear face only), or annulate. The radula is a single series of broadly arched (usually indented in the midline) **pectinate** teeth. Most species feed on anemones.

Aeolidiella drusilla (Aeolidiidae), Flinders Pier, Vic. John Chuk.

Anteaeolidiella lurana

Usually the white body has a series of orange, diamond-shaped patches with white centres along the midline of the back. The rhinophores are slightly ridged. The cylindrical cerata are white with a subapical orange or red band. A revision in 2014 showed that specimens from warm waters are pale to deep orange, with or without the diamond shapes along the back, and represent a different species, *Anteaeolidiella cacaotica*, which occurs as far south as Twofold Bay, NSW. Some specimens are entirely deep red.

Size: 15–20 mm, often smaller.

Habitat and range: under stones in intertidal pools to 15 m depth. Vic., Bass Strait and wide ranging in the Atlantic and Pacific. Previously misidentified as *Anteaolidiella foulisi*, a synonym of *Anteaeolidiella cacaotica* from NSW.

Baeolidia australis

Body plump, with wide tentacle-shaped anterior foot corners; usually mid-brown in colour, though may be much paler; always with whitish speckling. A large, rounded bluish spot outlined in white is present in front of the rhinophores between the oral tentacles. The cerata are broad, somewhat flattened and brown, reddish brown, greenish or bluish, with a narrow subapical red ring and sparse-to-dense cream spotting. The rhinophores are body coloured and finely papillate all around.

Size: to 50 mm, usually 25–35 mm.

Habitat and range: rare intertidally to 12 m depth. Southern Qld to southern WA, including Tas.; also NZ.

Baeolidia macleayi

A small, broad, flattened species, particularly common intertidally under rocks. Colour generally yellowy orange with much yellowish-cream spotting. Rhinophores irregularly roughened and yellowish above, orange below; cerata short and stout, usually with two yellowish-cream rings below the tip; foot broad with short tentacle-shaped anterior corners.

Size: 15–20 mm.

Habitat and range: intertidal under rocks, often in pairs, to 30 m depth. NSW to SA, including Tas.; also NZ.

Anteaeolidiella lurana (Aeolidiidae), Flinders Pier, Vic. John Chuk.

Baeolidia australis (Aeolidiidae), Blairgowrie Marina, Vic. John Chuk.

Baeolidia macleayi (Aeolidiidae), Flinders, West Head, Vic. John Chuk.

Burnaia helicochorda

This species has a slender, white (or rarely deep-orange) body and short tentacle-shaped foot corners. The rhinophores have a white stalk and dark-brown, rarely paler-golden, finely lamellate clavus or thickened upper part. The cerata are long, cylindrical and bluntly tipped; the digestive gland is typically dark green (rarely yellowy brown) and it zigzags back and forth within the cerata. A deep-orange or golden H-mark is present on the head, running from the oral tentacles to the rhinophores and transversely between the latter.

Size: 20–25 mm.

Habitat and range: intertidal to 10 m depth. Northern NSW to western SA, including Tas.; also NZ.

Cerberilla incola

A small, flattened aeolid with broad foot and long tentacle-shaped foot corners that drag rearward beside the animal as it crawls. Dark-brown or black rhinophores short and close together; oral tentacles very long; cerata flattened, short anteriorly, becoming longer and folded over the dorsum posteriorly, with a subapical black inverted 'U' on the dorsal side. Body colour yellowish cream with a black line on oral tentacles and along anterior edge of foot.

Size: 10–15 mm.

Habitat and range: 0–50 m, burrowing in sand or sandy mud seeking small anemones. NSW, Vic. and Tas.

Cerberilla sp. RB1

Similar to *Cerberilla incola, Cerberilla* sp. RB1 has two black lines on the dorsal side of each ceras; it too is a burrowing species.

Size: 27 mm.

Habitat and range: depth 5–30 m. Vic. and SA.

Burnaia helicochorda (Aeolidiidae), Kitty Miller Bay, Phillip Island, Vic. Leon Altoff.

Cerberilla incola (Aeolidiidae), Blairgowrie Marina, Vic. John Chuk.

Cerberilla sp. RB1 (Aeolidiidae), Blairgowrie Marina, Vic. John Chuk.

Family Embletoniidae

Small, flattened, slender nudibranchs, 3–40 mm in length in life. Head with smooth, short rhinophores set rather far apart and a rounded velum (replacing oral tentacles). Cerata usually squatly fusiform, the tip divided into two short branches, with the tip of each branch again divided into two short nodes. There appear to be no cnidosacs. The cerata stand singly along each side of the body. The foot is rounded in front. At most, four species belong to the family.

Embletonia gracilis

Body, velum and rhinophores white; cerata coloured according to the digestive gland: pale pink, buff, fawn or pale brown, with lighter tips.

SIZE: 5–8 mm, tropical forms much larger.

HABITAT AND RANGE: crawling on algae in intertidal pools, and to 127 m depth. Qld to Vic. and Tas.; tropical and temperate Indo-West Pacific.

Family Eubranchidae

Generally small aeolids with three teeth in each row of the radula. Animal similar in shape and size to tergipedid aeolids, but rather than laying back upon the body, in Bass Strait eubranchids, the cerata stand more or less upright and are often ornamented with two or three circlets of points or knobs. Eubranchids feed on hydroids.

At least 18 species of eubranchid aeolids are known to occur from Bass Strait, with only one named. Images of four unnamed species are included here. Brief information on those not included here can be found in Burn (2006).

Eubranchus rubeolus

The creamy-white body of this species has a series of big blood-red spots along each side and along the centreline of the body. The rhinophores and oral tentacles are blood red, as are the cerata.

SIZE: 8 mm.

HABITAT AND RANGE: intertidal to 18 m; among algae and the seagrass *Amphibolis antarctica*. Vic. and Tas.

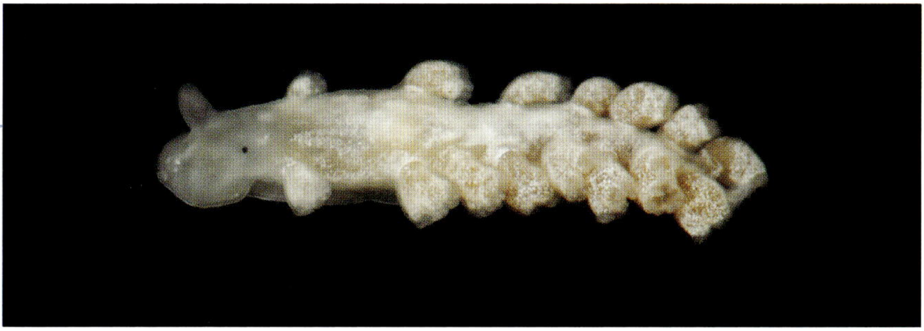

Embletonia gracilis (Embletoniidae), West Head, Flinders, Vic. John Chuk.

Eubranchus rubeolus (Eubranchidae), Cape Paterson, Vic. Platon Vafiadis.

Eubranchus sp. RB1

The white body of *Eubranchus* sp. RB1 often has patches or blotches of pale to dark brown along the sides and back; the rhinophores and oral tentacles have a brownish band at midlength, and the tall columnar cerata have three circlets of bluntly pointed knobs along their length. Importantly, the anal papilla opens on the right side of the back behind two rows of cerata. The smaller *Eubranchus* sp. RB2 (not figured) is very similar in shape and colours, but is separated by the position of the anal papilla behind three rows of cerata.

SIZE: usually 3–6 mm, exceptionally to 15 mm.

HABITAT AND RANGE: at times common on algae in lower intertidal pools to 10 m depth. Vic. and SA.

Eubranchus sp. RB3

This small, semitransparent species has a red-orange patch in front of the rhinophores, and the slightly bumpy cerata have black spots scattered over the digestive gland, a subapical bluish band, and the palest-yellow tip.

SIZE: 4 mm.

HABITAT AND RANGE: on algae in lower intertidal pools. Vic. Known only from six specimens recorded over 30 years.

Eubranchus sp. RB6

The white body is dorsally and laterally densely overlaid with red speckles, spots or blotches. The long, slender cerata are pale grey in colour, spirally twisted in shape, and studded with numerous low, hemispherical papillae.

SIZE: 12 mm.

HABITAT AND RANGE: among hydroids growing on algae at 8–10 m depth. Vic. and SA; also northern NZ. Five Australian specimens recorded over 40 years.

Eubranchus sp. RB1 (Eubranchidae), Point Lonsdale, Vic. Platon Vafiadis.

Eubranchus sp. RB3 (Eubranchidae), Point Lonsdale, Vic. Platon Vafiadis.

Eubranchus sp. RB6 (Eubranchidae), Popes Eye, Port Phillip Bay, Vic. Leon Altoff.

Eubranchus sp. RB9

Similar to *Eubranchus* sp. RB1 in body colouration and anal position; separated by the prominent white head, the tall slender cerata with three circlets of slightly raised points, and the dark-brown digestive gland visible in the cerata between the subapical vertical cream bars.

Size: 6–8 mm.

Habitat and range: on algae in lower intertidal pools, to 10 m depth. Vic.

Family Facelinidae

A large group of often long, slender species, mostly feeding upon hydroids, though one genus at least (Favorinus) eats the eggs of other nudibranchs. The rhinophores are often prominent, and may be smooth, annulate, swollen into bulbs, lamellate or papillate all round or on the rear face only. The cerata are cylindrical, slender, often curved or curled, with an apical cnidosac (except in Phyllodesmium, in which the cerata may be flattened, pustulose, ribbed or covered in points and there is no cnidosac). Many cerata may crowd each side of the body, though usually there are several ceratal groups visible, especially anteriorly. The anus opens among the cerata of the second group on the right side of the body (or exceptionally further rearward). The radula has a single series of narrow arched teeth, each with a single strong cusp and a number of smaller denticles either side.

Eubranchus sp. RB9 (Eubranchidae), Point Lonsdale, Vic. Platon Vafiadis.

Austraeolis ornata

A large species with semitransparent, orange body marked dorsally by a whitish stripe, the whole dotted with shining, almost transparent, blue spots, especially along the margins of the foot. Cerata are densely clustered, with a wide space between the anterior group and those behind; digestive gland usually dark, almost black, within the cerata; and the upper anterior surface is variously spotted and ringed with cream, gold and silver epidermal pigments. Oral tentacles long and thick; rhinophores stout, the upper part of each with 12 or more **annular** lamellae.

SIZE: to 40 mm, usually 15–25 mm.

HABITAT AND RANGE: common in intertidal pools and under rocks, to 100 m depth. Southern Qld to southern WA, including Tas. Juvenile and subadult specimens from shallow, subtidal waters (to 12 m depth) are often paler with a pink or red digestive gland within the cerata.

Cratena lineata

A small, slender aeolid with clear, whitish body overlaid by numerous, longitudinal, opaque white stripes and characteristic orange cheek-spots. Few cerata, distinctly grouped, digestive gland orange or brown, the tips white, more white lines along each ceras. Oral tentacles long and slender; the smooth rhinophores stand erect.

SIZE: to 20 mm, usually 10–15 mm.

HABITAT AND RANGE: at times common intertidally and subtidally to 18 m depth. Qld to WA and a periodic visitor to southern Australia; tropical and subtropical Indo-West Pacific.

Echinopsole breviceratae

Body slender, cerata short and tightly clustered with groups well separated. Oral tentacles long and stout, foot corners long and pointed. Rhinophores very distinctive, each with two strong, slightly undulate, annular lamellae. Colour of body deep pink to reddish brown, usually with many blue dots scattered along the margins of the midline cream stripe; smaller blue dots and specks along body sides and foot margins. Rhinophores and cerata pinkish fawn with yellowish speckling; oral tentacles and foot corners body-coloured with dense, cream patches.

SIZE: 30 mm.

HABITAT AND RANGE: intertidal pools/under rocks, to 10 m depth. Vic. and SA.

Austraeolis ornata (Facelinidae), Anderson Point, Portland, Vic. Platon Vafiadis.

Cratena lineata (Facelinidae), Kitty Miller Bay, Phillip Island, Vic. John Chuk.

Echinopsole breviceratae (Facelinidae), Point Leo, Vic. John Chuk.

Facelina hartleyi

A small aeolid with pink or white body marked by large opaque blue or white patches along the centreline. The cerata are almost continuous along each side of the body, their digestive gland deep red in colour, and the tips white; oral tentacles long and slender; rhinophores each with eight small annular lamellae in adult specimens.

SIZE: 12–15 mm.

HABITAT AND RANGE: common in intertidal pools and under rocks to 12 m depth. Vic., Tas. and SA.

Facelina newcombi

Very similar in shape to *F. hartleyi*. Very long and slender, the body fawn with brownish stripes on each side of the head. The cerata are short, with the anterior cluster separated from those behind. The digestive gland is deep to darkest brown, almost black, the tips white or cream; oral tentacles long and slender; rhinophores with 10–12 small annular lamellae.

SIZE: 20 mm, rarely longer.

HABITAT AND RANGE: intertidal pools and under rocks to 10 m depth. NSW and Vic.

Other *Facelina* species occur in the Bass Strait area. *Facelina* sp. RB3 closely resembles *F. hartleyi* and *F. newcombi* in size, shape and colour, but has posteriorly papillate rhinophores. *Facelina* sp. RB2 is purple or mauve with a yellow midline body stripe, yellow bands on the cerata, wrinkled rhinophores, and very wide and large tentacle-shaped foot corners. *Facelina* sp. RB1 has been recorded only twice (1981, 2011); it has yellow spots on the body, yellow stripes on the cerata, a blue line on the tail, and smooth rhinophores.

Facelina hartleyi (Facelinidae), Point Lonsdale, Vic. Platon Vafiadis.

Facelina newcombi (Facelinidae), Crawfish Rock, Western Port ,Vic. Andrew Newton.

Favorinus sp. RB1

A small aeolid of variable pigmentation (depending upon the colour of the seaslug eggs it has been eating), with a few dense clusters of small curled-in cerata each side of the body, and the rhinophores each bear three strong annulae, thickened below.

Size: to 12 mm.

Habitat and range: 0–10 m, often seen on and eating into the egg ribbons of *Pleurobranchus hilli*, *Aplysia* spp., *Doris cameroni*, *Trapania aureopunctata*, *Facelina hartleyi*, etc. Presently known only from Vic., SA and southern WA. Previously misidentified as *Favorinus pannuceus*.

Palisa sp. RB1

Opaque white line patterning upon a semitranslucent white body makes this species stand apart from all other aeolids in the Bass Strait area. The rhinophores are posteriorly strongly papillate, the cerata are long and curled in towards the midline, and the tentacle-shaped foot corners are long and narrow.

Size: to 15 mm.

Habitat and range: 0–3 m, crawling on brown algae intertidally and subtidally. Vic.

Phyllodesmium macphersonae

Unusual among southern Australian aeolid nudibranchs in that the body is deep grey from the close network of digestive gland ducts lined with zooxanthellae. The long, curved cerata are few in number, and the groups are distinct and deep blue in colour with lighter-yellow tips. There are no cnidosacs in the ceratal tips. The oral tentacles are long; the rhinophores are smooth, each with a grey base and yellow or cream upper half.

Size: 20–35 mm.

Habitat and range: intertidal pools and under rocks, to 13 m depth. All states; tropical and temperate Indo-West Pacific.

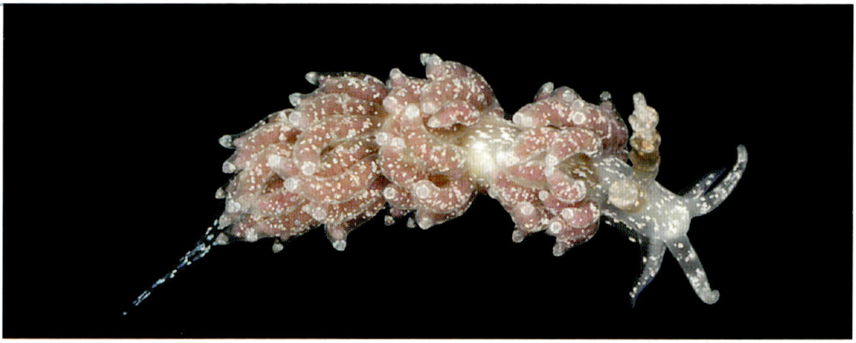

Favorinus sp. RB1 (Facelinidae), Portsea Pier, Vic. John Chuk.

Palisa sp. RB1 (Facelinidae), Sunderland Bay, Phillip Island, Vic. Platon Vafiadis.

Phyllodesmium macphersonae (Facelinidae), Point Lonsdale, Vic. Platon Vafiadis.

Phyllodesmium serratum

This species has a semitranslucent white body, oral tentacles and rhinophores. A white line is generally present along the midline of the back. The cerata are long, slender and curved or twisted. They are transparent, leaving the winding orange, pink or reddish digestive gland clearly visible. There are no cnidosacs in the ceratal tips.

SIZE: 15–40 mm.

HABITAT AND RANGE: intertidal pools and under rocks to 20 m depth. All states; tropical and temperate Indo-West Pacific.

Family Fionidae

This family comprises one genus with one cosmopolitan species. The rhinophores and oral tentacles are short and tapering. The cerata are very distinctive, with a prominent full-length, wavy crest projecting from the inner side.

Fiona pinnata

Body pink or grey, cerata bluish, the projecting crest white along the edge.

SIZE: 10–25 mm.

HABITAT AND RANGE: living among and eating goose barnacles growing on flotsam in the open ocean; often washed ashore on cuttle bones after storms. NSW, Vic. and Tas.; cosmopolitan.

Phyllodesmium serratum (Facelinidae), Mornington Pier, Vic. John Chuk.

Fiona pinnata (Fionidae), Eden, NSW. Bill Rudman, Australian Museum.

Family Flabellinidae

Aeolid nudibranchs in which a slender notal brim, or traces of it, is often present along each side of the body. The cerata are always dorsal to the notal brim, with the anus below. The rhinophores can be simple and smooth, papillate all round (or only on the rear face), or annulate. The radula is comprised of a denticulate central tooth with one (rarely two) smooth or denticulate lateral tooth (teeth) each side. All species feed on hydroids.

Flabellina poenicia

A small, slender aeolid with white or pale-pink body, long slender cerata, each with a vivid-pink, dark-red or light-brown digestive gland and large white cnidosac, and sparse papillae on the rear face of the rhinophores. The jaws are visible externally as a vivid-pink or purple patch on either side of the head. The anterior foot has tentacle-like corners.

Size: 15 mm, but usually smaller.

Habitat and range: intertidally crawling on algae or under rocks, to 55 m depth. NSW, Vic., Tas. and SA.

Flabellina sp. RB1 and *Flabellina* sp. RB2

Both species have a purplish body with darker-purple oral tentacles, rhinophores and tail tips, orange or reddish cerata, and transversely wrinkled rhinophores. They are separated by the shape of the anterior edge of the foot: *Flabellina* sp. RB1 has tentacle-like corners; *Flabellina* sp. RB2 has a broadly rounded, wider foot without corners.

Size: 10–15 mm, rarely to 25 mm.

Habitat and range: 0–10 m, sometimes associated with the hydroid *Eudendrium generale*, oft-times crawling on algae in deep rock pools. *Flabellina* sp. RB1 is presently known only from southern NSW, Vic. and SA, *Flabellina* sp. RB2 from both Vic. and Tas.

Flabellina poenicia (Flabellinidae), Portsea Pier, Vic. John Chuk.

Flabellina sp. RB1 (Flabellinidae), Popes Eye, Port Phillip Bay, Vic. Platon Vafiadis.

Flabellina sp. RB2 (Flabellinidae), Aireys Inlet, Vic. Platon Vafiadis.

Flabellina sp. RB4

A slender species with relatively few cerata along each side of the body, tall rhinophores with dense papillae on the posterior side of the upper half, and long thin tentacle-like foot corners; body colour deep mauve; cerata bluish mauve with a red subapical band and small white tip.

SIZE: to 25 mm, but usually smaller.

HABITAT AND RANGE: 0–10 m, sometimes associated with the hydroid *Eudendrium generale*. Presently known only from central Vic. (Hitherto confused with *Flabellina rubrolineata* from warmer waters, but the latter species has a much paler, whitish body with a line of red pigment along the midline of the dorsum and another each side of the body below the cerata.)

Tularia bractea

The colours of this species seem to sparkle with brightness. The body is opaque white overlaid with specks of bright bluish white on the head and along the back, with dashes of the same colour along the lateral bases of the cerata. The digestive gland within the cerata may be yellowish green, green, reddish brown, or burnished orange, each ceras sparkling with golden dots. The rhinophores are smooth or a little wrinkled. The anterior foot has long, tentacle-like corners. The cerata stand nearly upright on projecting flanges along each side of the body.

SIZE: 15–25 mm.

HABITAT AND RANGE: intertidal to 15 m depth; under stones or crawling on algae. NSW, Vic., Tas. and SA; also NZ; periodically common.

Flabellina sp. RB4 (Flabellinidae), Aireys Inlet, Vic. Platon Vafiadis.

Tularia bractea (Flabellinidae), Point Danger, Torquay, Vic. Platon Vafiadis.

Family Glaucidae

A family of one genus comprised of five species. Glaucid aeolids are adapted to a pelagic life on the open ocean. There they spend their lives floating upside down at the surface of the sea, helped by an ingested bubble of air. Animals of both species are bicoloured: dark and darkest blue on the ventral side, which faces the light; silvery white on the dorsal surface, which faces the abyssal depths. The cerata arise from stout arms or peduncles that project from the body. The cerata are used to 'row' animals about as they seek their prey (the floating cnidarians *Physalia, Velella* and *Porpita*).

Glaucus atlanticus

Cerata long and slender, set in a single row upon each peduncle; foot with two dark stripes.

SIZE: 20 mm, but usually smaller.

HABITAT AND RANGE: pelagic, washed ashore after storms. Eastern Vic. and north-eastern Tas.; cosmopolitan.

Glaucus marginatus

Cerata shorter and crowded, set in a double row upon each peduncle; foot with an entirely dark sole. Recent research indicates *G. marginatus* comprises at least four cryptic sibling species. *Glaucus marginatus* is now restricted to tropical and subtropical Indo-Pacific regions. The species occurring in the southern Pacific, including eastern Bass Strait, has been described as a new species, *Glaucus bennettae*.

SIZE: 10 mm, often smaller.

HABITAT AND RANGE: pelagic, washed ashore after storms; less common than *Glaucus atlanticus*. Indo-Pacific.

Glaucus atlanticus (Glaucidae), Collaroy, Sydney, NSW. Bill Rudman, Australian Museum.

Glaucus marginatus (Glaucidae), Collaroy, Sydney, NSW. Bill Rudman, Australian Museum.

Family Tergipedidae

A large family made up of generally small and often colourful aeolid nudibranchs. All species eat hydroids or the flesh of corals, and are usually difficult to see when upon or among their food. The body is usually slender and the anterior foot rounded or with tentacle-shaped foot corners. The oral tentacles are long and slender, and the rhinophores mostly smooth, though species with annulate, spirally twisted, lamellate, studded or papillate rhinophores occur locally. The cerata vary considerably in number from a series of single cerata well spaced along each side of the body to densely crowded cerata so close that little of the body is visible. The radula is a series of single, denticulate teeth. The anus opens on the right side between the first and second clusters of cerata. A total of 30 tergipedid species are known to occur within the Bass Strait area, of which only seven are named. Some are abundant intertidally and in the shallow subtidal zone; some have been sighted only once in 55 years. Images of nine of the unnamed species are included here. (Details of some of the remainder can be found in Burn (2006).) Several species of *Trinchesia* have been recently moved to the genus *Cuthona*; however, until a global revision of the family is available, local species are here retained in *Trinchesia*.

Tergipes sp. RB1

A very small, slender aeolid with only three or four long, spindly cerata along each side of the body; oral tentacles and rhinophores long and smooth; body dull-cream with a blush of red or reddish brown on the top and sides of the head; digestive gland in cerata brownish.

SIZE: 4–5 mm, rarely 7–8 mm.

HABITAT AND RANGE: 0–18 m, usually crawling on brown algae in intertidal pools; observed once in vast numbers feeding on the hydroid *Silicularia rosea* growing on the brown alga *Phyllospora*, Pirates Bay, Tas. (November 2007). Vic., Tas., SA and south WA.

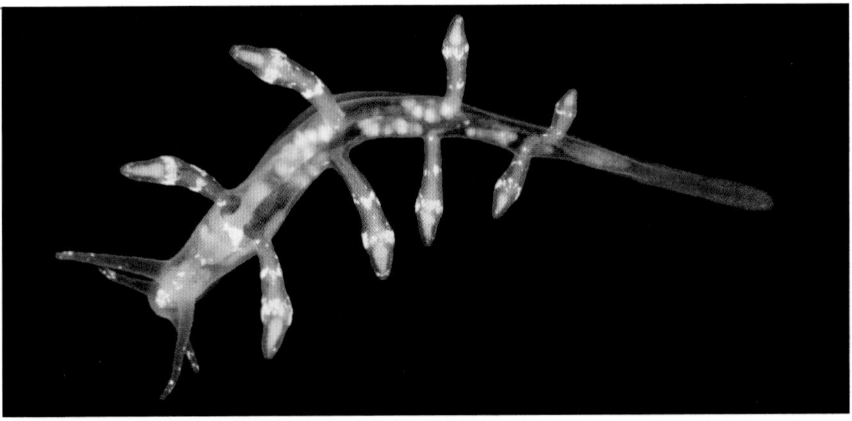

Tergipes sp. RB1 (Tergipedidae), Popes Eye, Port Phillip Bay, Vic. Leon Altoff.

Trinchesia anulata

White all over with relatively few, long, strongly curved cerata in six groups along each side of the body. The stout, tapering rhinophores each have five or six annulate lamellae.

SIZE: 6–10 mm.

HABITAT AND RANGE: uncommon, intertidal on algae in pools, to 20 m depth. Vic.; tropical and temperate Indo-West Pacific.

Trinchesia catachroma

The head and lower part of the rhinophores are always suffused with an orange-red colour. The tips of the rhinophores and oral tentacles are opaque white or yellowish. The body is long and slender. The cerata are stoutly fusiform, varying in colour (brown, green or blue), with bright yellowish tips. The rhinophores are smooth.

SIZE: 8–15 mm.

HABITAT AND RANGE: common intertidally crawling on algae in pools, to 12 m depth. Vic. and SA.

Trinchesia kuiterorum

This species is amazingly camouflaged, the upper end of each ceras with two circlets of long slender tentacles, simulating the hydroid upon which it feeds. The entire animal is pale fawn, with a reddish patch indicating the muscular pharynx within the head.

SIZE: 8 mm.

HABITAT AND RANGE: subtidal, 1.5–10 m depth; among and eating the solitary athecate (i.e. order Anthomedusa) hydroid *Zyzzyzus spongicola* growing on orange or brown sponges. NSW and here newly recorded from the marine fauna of Vic. (southern Port Phillip Bay).

Trinchesia anulata (Tergipedidae), Point Addis, Vic. Leon Altoff.

Trinchesia catachroma (Tergipedidae), Point Lonsdale, Vic. Platon Vafiadis.

Trinchesia kuiterorum (Tergipedidae), Blairgowrie Marina, Vic. John Chuk.

Trinchesia ornata

The strong yellowish-orange body is unlike that of any other Bass Strait *Trinchesia*. The tips of the oral tentacles, the rhinophores and the tail are yellow; the short cerata have a dark-greenish or brownish digestive gland overlaid with blue pigment, each surmounted by a yellow tip.

Size: 8–12 mm.

Habitat and range: 0–10 m locally, crawling on brown algae and seagrass in low intertidal areas. An occasional visitor from warmer waters. Qld. and Vic.; Japan and southern Africa.

Trinchesia sororum

The digestive gland in the cerata is a rich, red colour, and the white cnidosacs are relatively large. The body is white, and the oral tentacles have a single pale-purple band, whereas the rhinophores have two pale-purple bands.

Size: 5 mm.

Habitat and range: intertidal, crawling on brown algae in lower intertidal rock pools. Vic. and SA.

Trinchesia thelmae

A beautiful, delicate pink species characterised by the presence of cup-like semi-annulae on the rear face of the rhinophores, and by the cerata with reddish-brown digestive gland overlaid by longitudinal white lines and yellowish tip.

Size: 8–20 mm.

Habitat and range: uncommon, intertidally crawling on algae in pools, to 10 m depth. Vic. and SA.

Trinchesia ornata (Tergipedidae), San Remo, Vic. John Chuk.

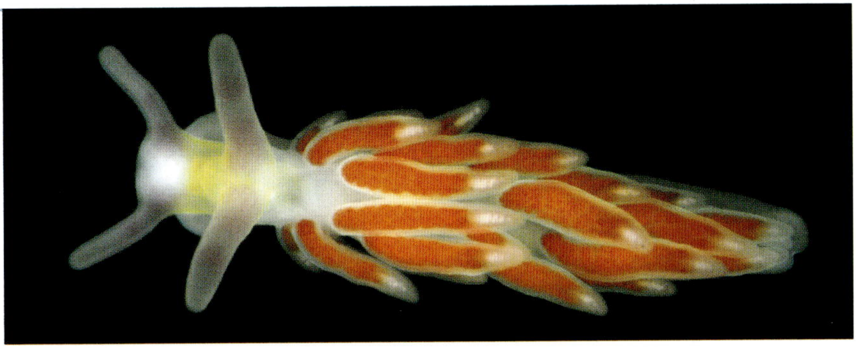

Trinchesia sororum (Tergipedidae), Point Lonsdale, Vic. Platon Vafiadis.

Trinchesia thelmae (Tergipedidae), Blairgowrie Marina, Vic. John Chuk.

Trinchesia viridiana

The body, oral tentacles and rhinophores of this aeolid are yellowish cream. The rhinophores are smooth and short. The cerata are rather few, the clusters well separated, the digestive gland brownish green to dark green, and within the tip of each is a small, shining cream or yellowish cnidosac.

SIZE: to 15 mm, but usually less than half this size.

HABITAT AND RANGE: 0–10 m, crawling on brown algae in low intertidal areas. Vic. and Tas.

Trinchesia sp. RB1

A most distinctive small species, easily recognised by the blood-red spots or patches on the sides and back of the semitranslucent, white body, the bright-white stripe on the tail, the opaque yellowish or cream cerata, and the white-tipped rhinophores and oral tentacles.

SIZE: 6 mm.

HABITAT AND RANGE: on algae in lower intertidal pools. Vic. A rare species, recorded about six times in 50 years.

Trinchesia viridiana (Tergipedidae), Cat Bay, Phillip Island, Vic. Leon Altoff.

Trinchesia sp. RB1 (Tergipedidae), Point Danger, Torquay, Vic. Platon Vafiadis.

Trinchesia sp. RB2

The dull-yellowish cerata of this species have a bright-yellow tip, below which is a blue spot on the anterior face. The upper two-thirds of the rhinophores are pinkish or opaque white, while the oral tentacles are blue. Subtidal specimens are sometimes more brightly pigmented.

Size: 5–7 mm.

Habitat and range: at times common, crawling on algae and the seagrass *Amphibolis antarctica* in lower intertidal pools, to 4–5 m depth. Vic.

Trinchesia sp. RB3

A small, flattened species, the cerata with a slight bend and ceras tips projecting beyond body at 30–40°. Digestive gland in cerata dark brown basally and light brown to fawn distally, with the middle section obscured by an oblique, opaque band; rhinophores and oral tentacles with a midlength, orange band.

Size: 4–5 mm.

Habitat and range: at times very common on the seagrass *Amphibolis antarctica* from the low intertidal to 10 m depth. Vic., Tas. and SA. This species hunts for hydroids that grow on the close-pressed leaves of *Amphibolis antarctica*.

Trinchesia sp. RB5

The dark-brown or black speckling of the clearly visible digestive duct within the body and digestive gland in the cerata distinguish this species. The cerata are rather stout and are often recurved. The rhinophores and oral tentacles are longer than the cerata and have a wide, brownish-orange band at midlength. Body very often densely speckled with silvery-white dots.

Size: 5–7 mm.

Habitat and range: on algae in lowest intertidal pools, to 18 m depth. Vic., Tas. and SA.

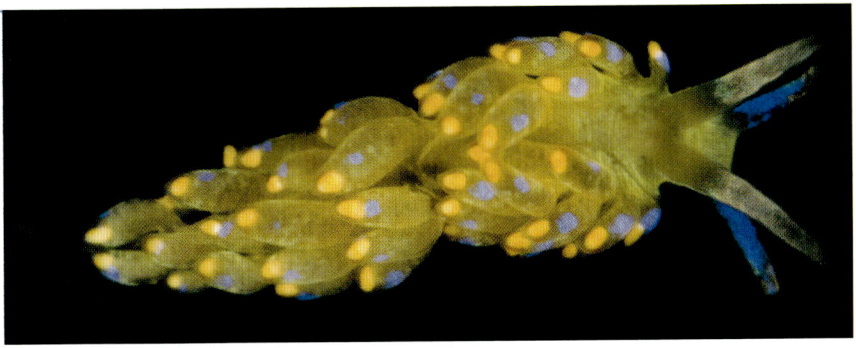

Trinchesia sp. RB2 (Tergipedidae), Point Addis, Vic. Leon Altoff.

Trinchesia sp. RB3 (Tergipedidae), Mushroom Reef, Flinders, Western Port, Vic. Leon Altoff.

Trinchesia sp. RB5 (Tergipedidae), Blairgowrie, Port Phillip Bay, Vic. Platon Vafiadis.

Trinchesia sp. RB7

Cerata opaque white with subapical bluish band, below which there is a bright-yellow spot on the anterior face. Rhinophores and oral tentacles semi-translucent with a midlength, opaque white band.

Size: 4–5 mm.

Habitat and range: on algae in lower intertidal pools, to 3 m depth. Vic. A rare species recorded only three or four times in 30 years.

Trinchesia sp. RB10

Head orange or reddish, and rhinophores and oral tentacles yellowish, as in *Trinchesia catachroma*, but overall shorter and broader than that species. Cerata fat basally, each with three or four rings of yellow pigment.

Size: to 8 mm.

Habitat and range: on algae in lower intertidal pools, to 10 m depth. Vic. Found together with *Trinchesia catachroma*, but much less common.

Trinchesia sp. RB7 (Tergipedidae), Point Lonsdale, Vic. Platon Vafiadis.

Trinchesia sp. RB10 (Tergipedidae), Point Lonsdale, Vic. Platon Vafiadis.

Trinchesia sp. RB11

The largest species of the genus *Trinchesia* in Victorian waters. Somewhat similar in colouration to *Trinchesia catachroma* and *Trinchesia* sp. RB10, this species has a blue subapical band on the cerata, as does *Trinchesia* sp. RB10. The cerata are numerous, close-set and completely obscure the body. The rhinophores are very distinctively spirally twisted.

SIZE: 18 mm, but usually smaller.

HABITAT AND RANGE: on algae in lower intertidal pools, to 12 m depth. Vic.

Trinchesia sp. RB14

Similar in colour and pattern to *Facelina* sp. RB2, but lacks the long tentacle-shaped foot corners of that species. Body pale mauve; tips of rhinophores, oral tentacles and tail reddish purple; cerata with a wide opaque band just above midlength, above which there is a narrow bright-yellow ring.

SIZE: 8 mm.

HABITAT AND RANGE: on algae in lower intertidal pools, to 10 m depth. Vic. and SA.

Trinchesia sp. RB11 (Tergipedidae), Popes Eye, Port Phillip Bay, Vic. Platon Vafiadis.

Trinchesia sp. RB14 (Tergipedidae), Point Danger, Torquay, Vic. Platon Vafiadis.

APPENDIX
FURTHER INFORMATION

This guide has been written as a practical introduction to nudibranchs and related molluscs of the Bass Strait region. Those keen to learn more can consult the publications and websites listed below. Please note, this is only a small selection of the many important text and web resources available.

PUBLICATIONS

Beesley, P.L., Ross, G.J.B. and Wells, A. (Eds). 1998. *Mollusca: The Southern Synthesis. Fauna of Australia Volume 5*. CSIRO Publishing: Melbourne.

Behrens, D.W. 2005. *Nudibranch Behavior.* New World Publications Inc.: Jacksonville.

Burn, R. 1989. Opisthobranchs (Subclass Opisthobranchia). *In:* Shepherd, S.A. and Thomas, I.M. (Eds). *Marine Invertebrates of Southern Australia. Part II*, pp. 725–788. South Australian Government Printing Division: Adelaide.

Burn, R. 1998. Orders Acochlidia, Rhodopemorpha, Sacoglossa. *In:* Beesley, P.L., Ross, G.J.B. and Wells, A. (Eds). *Mollusca: the Southern Synthesis. Fauna of Australia Volume 5*, pp. 959–974. CSIRO Publishing: Melbourne.

Burn, R. 2006. A checklist and bibliography of the Opisthobranchia (Mollusca: Gastropoda) of Victoria and the Bass Strait area, south-eastern Australia. *Museum Victoria Science Reports*, 10: 1–42. [http://museumvictoria.com.au/sciencereports/]

Burn, R. and Thompson, T.E. 1998. Order Cephalaspidea. *In:* Beesley, P.L., Ross, G.J.B. and Wells, A. (Eds). *Mollusca: The Southern Synthesis. Fauna of Australia Volume 5*, pp. 943–959. CSIRO Publishing: Melbourne.

Cobb, G. and Willan, R.C. 2006. *Undersea Jewels: A Colour Guide to Nudibranchs.* ABRS: Canberra.

Coleman, N. 2001. *1001 Nudibranchs – Catalogue of Indo-Pacific Sea Slugs.* Neville Coleman's Underwater Geographic Pty Ltd: Springwood, Queensland.

Coleman, N. 2008. *Nudibranchs Encyclopedia. Catalogue of Asia/Indo-Pacific Sea Slugs.* Neville Coleman's Underwater Geographic Pty Ltd: Springwood, Queensland.

Debelius, H. and Kuiter, R.H. 2007. *Nudibranchs of the World.* IKAN-Unterwasserarchiv: Frankfurt.

Edgar, G.J. 2000. *Australian Marine Life, the Plants and Animals of Temperate Waters.* Revised edition. Reed Books: Kew, Australia.

Gosliner, T.M., Behrens, D.W. and Valdés, A. 2008. *Indo-Pacific Nudibranchs and Sea Slugs.* Sea Challengers Natural History Books: Gig Harbour, Washington.

Gowlett-Holmes, K. 2008. *A Field Guide to the Marine Invertebrates of South Australia.*

Notomares: Sandy Bay, Tasmania.

Marshall, J.G. and Willan, R.C. 1999. *Nudibranchs of Heron Island, Great Barrier Reef. A Survey of the Opisthobranchia (Sea Slugs) of Heron and Wistari Reefs.* Backhuys: Leiden.

Phillips, D.A.B., Handreck, C.P., Bock, P.E., Burn, R., Smith, B.J. and Staples, D.A. (Eds). 1984. *Coastal Invertebrates of Victoria. An Atlas of Selected Species.* Marine Research Group of Victoria and Museum of Victoria: Melbourne.

Schrödl, M., Klussmann-Kolb, A. and Wilson, N.G. 2011. Bye bye "Opisthobranchia"! A review on the contribution of mesopsammic sea slugs to euthyneuran systematics. *Thalassas*, 27: 101–112.

Wells, F.E. and Bryce, C.W. 1993. *Sea Slugs of Western Australia.* Western Australian Museum: Perth.

Willan, R.C. and Coleman, N. 1984. *Nudibranchs of Australasia.* Australasian Marine Photographic Index: Caringbah, New South Wales.

WEBSITES

Australian Faunal Directory – checklists of Australian species:
http://www.environment.gov.au/biodiversity/abrs/online-resources/fauna/afd/home

Atlas of Living Australia – species lists, collections, maps and images:
http://www.ala.org.au/

Museum Victoria Collections – a selection of collection specimens including species identification and distribution data:
http://collections.museumvictoria.com.au/

Nudibranchs.com.au – by Gary Cobb and David Mullins:
http://www.nudibranch.com.au/index.html

Nudi Pixel – nudibranch photographic identification tool:
http://www.nudipixel.net/

World Register of Marine Species (WoRMS) – worldwide checklist of marine species:
http://www.marinespecies.org/

Sea Slug Forum – by Bill Rudman and the Australian Museum:
http://www.seaslugforum.net/

The Port Phillip Bay Taxonomic Toolkit – information, data, images and tools to help identify, document and monitor the marine animal life of Port Phillip Bay:
http://portphillipmarinelife.net.au/

ACKNOWLEDGEMENTS

Funding for the Museum Victoria Field Guides to Marine Life series has come from the Australian Government through a grant from its Natural Heritage Trust. The grant was facilitated by the Port Phillip and Westernport Catchment Management Authority. Publication of the series is funded in part by the Victorian Coastal Council and Parks Victoria. This title within the series also had direct funding from the Norman Wettenhall Foundation, for which the author and editors are especially grateful.

This guide condenses more than 60 years of amateur research by the writer on the nudibranchs and related molluscs of Victoria and nearby areas. The grant from the Norman Wettenhall Foundation to Museum Victoria provided much-needed help to create the manuscripts for the initial checklist (published in 2006) and also this field guide. Present and past staff of the Marine Biology Department of Museum Victoria, in particular Liz Greaves, Melanie Mackenzie, Chris Rowley and Drs Robin Wilson and Anna McCallum, have facilitated access to specimens, literature, typing, imaging and editing.

For comments, all of which helped improve the manuscript, thanks go to John Chuk, Platon Vafiadis, Ken Bell, Bill Rudman, Jan Watson and Richard Willan. Richard Willan provided detailed and constructive comments on the penultimate version of the manuscript, and the vast majority of his suggestions have been gratefully adopted. Jeanette Birtles provided expert editorial guidance.

Lastly and very importantly, my most sincere and grateful thanks go to all those too numerous to name, who over a lifetime's interest have helped by securing specimens at near and far locations, in shallow and deeper waters, in fair weather and foul. Without you, this guide would not have been possible.

PHOTOGRAPHS AND ILLUSTRATIONS CREDITS

Images are essential to this guide, and for the use of these the writer is most grateful to many colleagues, friends and associates. In particular, thanks go to John Chuk, Leon Altoff and Platon Vafiadis for the care and attention given over many years to photographing animals, sometimes at very short notice. Thanks are also offered to Sylvia Buchanan, the late Neville Coleman, Andrew Newton, Trevor McMurrich, Glenys Greenwood, Joan Hales, Don Love, Jerome Mallefet, Mark Norman, Bill Rudman and Richard Willan, each of whom contributed images and specimens, as well as to all those whose contributions are noted throughout the text.

ABOUT THE AUTHOR

Robert Burn is an internationally respected authority on the marine group to which nudibranchs and related molluscs belong. Imbued with a love of these animals since childhood, he has over 60 years of experience collecting, observing and studying nudibranchs in the field and under the microscope. Robert's expertise is acknowledged through his honorary status with the Marine Biology section of Museum Victoria, the Malacological Society of Australasia, and the Field Naturalists Club of Victoria. He has been an Honorary Associate with Museum Victoria since 1962 and has been a member of the Malacological Society since early in 1952, the year of its foundation. Alongside his 'day job' as a third generation building contractor in Geelong, Victoria, Robert has authored more than 100 publications on nudibranchs and other molluscs, including sections for *Coastal Invertebrates of Victoria: An Atlas of Selected Species* (1984), *Marine Invertebrates of Southern Australia* (1989), and *Fauna of Australia, Mollusca: The Southern Synthesis* (1998). He has described more than 90 species of nudibranchs and related molluscs, the vast majority from the coastline of Victoria. Many more species are known and await description.

GLOSSARY

Aeolidiform resembling **aeolid nudibranchs**, i.e. with **cerata** along each side of the body

Anastomosing forming a network

Annular ring-shaped

Annulate with two to many ring-like projections along the length

Aperture opening, e.g. of shell or of reproductive system

Ascidian an animal belonging to the sea squirt class Ascidiacea

Bifid forked; split into two parts

Bilobed with two lobes

Bivalves molluscs with a shell made of two hinged pieces; Bivalvia (clams, mussels, scallops etc.) are a separate class from Gastropoda, but there are a number of bivalved **gastropods** (Family Juliidae)

Branchiae gills

Bulliform bubble-shaped

Cerata (sing. ceras) clustered tubular outgrowths along each side of the body in some **nudibranchs** and **sacoglossans**; each ceras acts as a gill and contains a branch of the digestive gland

Ceratiform shaped like a **ceras**

Cerianthid anemones tube-dwelling anemones

Chemosensory capable of detecting chemicals, e.g. describes receptors that can detect odours or tastes i.e. the **rhinophores**

Clavus top part of **rhinophore**, often enlarged and decorated with **lamellae** or **papillae**

Cnidarians members of the **phylum** Cnidaria, containing jellyfish, anemones, corals and hydroids

Cnidosac a sac in the tip of the **cerata** of **aeolids** for storage of functional stinging cells (**nematocysts**) ingested while feeding upon **cnidarians**

Columella column at the centre of a coiled snail shell

Co-marginal parallel to the margin

Concretion accumulated substance forming a hard mass

Confluent merging

Cryptic being difficult to see due to body camouflage against the background e.g. **host-food**

Cryptic species species which are difficult or impossible to distinguish using morphological characters alone

Cyanobacteria photosynthetic bacteria

Dendritic branching, like a tree

Denticles smaller cusps or points either side of main (median) cusp of each **radula** tooth

Dorsum/Notum upper surface of the **mantle**

Ductules small ducts of the digestive gland

Elongate oval extended lengthwise and rounded each end

Enrolled cylindrical with a lateral groove

Epifaunal living on the surface of the **substrate**

Eviscerate remove the inner organs from

Extensile setae extendable bristles

Faecal waste waste product from digestion

Flagellum long slender posterior appendage of some bubble shells

Foot muscular region on the underside of a **mollusc** used for locomotion

Fouling community a group of interacting species living on non-natural surfaces such as ship hulls

Fusiform spindle-shaped

Gastral plates hard, calcareous or chitinous plates, often distinctively shaped; present in the gizzard of some bubble shells, sea hares and **nudibranchs** for grinding ingested food

Gastropod a class of **molluscs** including snails and slugs; if a shell is present, there is one shell rather than two shells

Hancock's organs paired **chemoreceptors** in the groove between the foot and the headshield in bubble shells

Headshield a widening of the head used to push through a sandy **substrate** without allowing the sand into the **mantle** cavity; found in the suborder Cephalaspidea (**headshield** slugs and bubble shells)

Hermaphroditic having a reproductive system that is complete and successful as both male and female

Host-food the organism upon which a nudibranch is found and which it typically eats

Hyaline having a glassy appearance

Hydroids animals of the class Hydrozoa in the **phylum** Cnidaria; resemble inverted jellyfish attached to a **substrate** and form the food of a number of nudibranch **species**

Hyponotum under-surface of the **mantle** or hyponotal overhang

Infaunal living or burrowing within the **substrate**

Lamellate with few to many projecting leaf-like projections, more or less horizontal, angled or vertical

Lappets muscular plates or protruberances around the mouth

Mantle body wall covering the upper surface of a sea slug; secretes any shell present, and forms the **mantle** cavity used for **respiration** and excretion and which contains the **visceral mass**

Molluscs a **phylum** of soft bodied invertebrate animals with a **mantle** cavity, a **radula** and a nervous system; external shell usually present

Nematocysts stinging cells of **hydroids**, jellyfish, sea anemones and other **cnidarians**

Notal brim remnant of or greatly reduced **notum**, generally visible as a low ridge or line of projecting points or dendritic appendages separating the body sides from dorsal surface (also referred to as pallial line, pallial brim, pallial points).

Notal margin edge of the **notum**

Notum usually broadly oval dorsal surface of various sea slugs, overhanging **visceral mass** all round; notal brim is edge of notum (see also **mantle**).

Nudibranchs 'naked-gilled' **gastropods**; an (often colourful) group of sea slugs that shed their shells during metamorphosis into the adult form; note that some other types of sea slugs are not closely related to **nudibranchs**

Ocelli eye-like spots, coloured (most often shining blue) and surrounded by a ring (or rings) of different colours

Operculum horny or calcareous plate, partially or completely closing the **aperture** of the **gastropod** shell on retraction of the animal within

Opisthobranchs a group of marine

gastropods distinguished by a single gill to the rear and right of the heart; currently under **taxonomic** revision

Oral cowl enlarged **velum** forming a hood-shaped structure over the mouth that can be projected over prey like a net (only in family Tethydidae)

Oral tentacles pair of tactile organs located beside or above the mouth of some slugs

Pallial pertaining to the **mantle**

Pallial brim see **notal brim**

Pallial line see **notal brim**

Pallial points see **notal brim**

Papilla (pl. papillae) small finger-like projection(s)

Papillate with one or more papillae

Parapodia plural of **parapodium**

Parapodium thin, lateral wing-like expansion of the **foot** in some slugs; protects the shell or used for swimming

Pectinate comb-like

Pericardial ridge raised rearward extension of the **pericardium** (in **sacoglossans**)

Pericardium a cavity around the heart and the mollusc equivalent of kidneys; often raised like a blister in **sacoglossans** and aeolid nudibranchs

Periostracum thin 'skin' that forms the outer layer of the shell of a **mollusc**

Pharynx muscular organ within the forepart of the body, containing the jaws and **radula**

Phylum a major **taxonomic** group of organisms, encompassing a group of classes; classes are groups of orders, which in turn are groups of families; families are groups of genera, which in turn are groups of **species**

Protoconch the tip of a coiled snail shell, secreted by the embryo before hatching or by the **veliger** before metamorphosis

Pulmonates air-breathing slugs

Punctae fine pores or pits

Punctate interrupted periodically with **punctae**

Pustules small, liquid-filled bumps

Pyriform pear-shaped

Quadrately rectangularly shaped in general outline

Radula a ribbon of chitinous teeth in regular longitudinal and transverse rows, located in the **pharynx**

Respiration gaseous interchange between an organism and its surrounding medium; includes gas exchange via a single gill supplied with a water current in the **mantle** cavity in many **gastropods**; in some **nudibranchs** the gill has been replaced by **cerata**

Rhachidian tooth differentiated central tooth in each transverse row of radular teeth

Rhinophores well-developed, paired **chemosensory tentacles** on the head of **nudibranchs**, **sacoglossans**, sea hares, **umbraculids** and side-gilled slugs

Sacoglossans sap-sucking sea slugs and sea snails from the order Sacoglossa; they mainly feed on marine algae

Secondary papilla circlet of small rounded or pointed projections ornamenting the **cerata**, especially in the family Dotidae

SEM Scanning Electron Microscope

Serpulid polychaetes marine worms of the family Serpulidae, which build and live in largely calcium carbonate ('lime') tubes

Sheath flange at or surrounding rhinophoral base

Species specifies the 'type' or 'kind' of an organism; defined as the largest group of similar organisms capable of interbreeding to produce fertile offspring

Spermatophores packets of sperm

Striae small linear marks, grooves or channels

Substrate surface in or on which an organism resides, e.g. rock, sand, mud

Sutural slit incision between the **whorls** of the shell

Symbiotic having a long-term, close relationship between two or more different **species**; may or may not benefit one or both **species**

Systematics the study of the diversity of organisms and their evolutionary relationships, and the development of a useful classification system

Taxonomic to do with the science of describing and classifying organisms into a hierarchy of categories, e.g. **phylum**

Teleoconch makes up the main part of the snail shell (other than the **protoconch**) and is secreted by the adult snail mantle

Tentacles oral tentacles (front pair) and **rhinophores** (rear pair); both may function as organs of touch and 'smell' (detecting molecules dissolved in water, similar to the sense of smell in air-breathing animals)

Truncate bluntly and abruptly

Tubercle small, rounded process projecting from the surface

Umbo the apex or tip of a valve of a **bivalve**, also called a 'beak'

Umbraculid organisms in the **gastropod** order Umbraculida, e.g. false limpets

Velar of the **velum**

Veliger uniquely **molluscan** larva, characterised by the ciliated, **bilobed velum** used for feeding and swimming

Velum anteriorly rounded, fleshy projection over the mouth of a **veliger** used for swimming and feeding; in adults front edge often with projecting smooth or **dendritic papillae**

Visceral mass/hump the soft non-muscular rear part of a bubble shell or sea hare in which most of its organs reside; located beside the **mantle** cavity

Whorl a spiral or circular pattern, e.g. one of the coils in a **gastropod** shell

Zooxanthellae symbiotic unicellular algae living within an animal

INDEX OF SCIENTIFIC NAMES

Other titles in the series

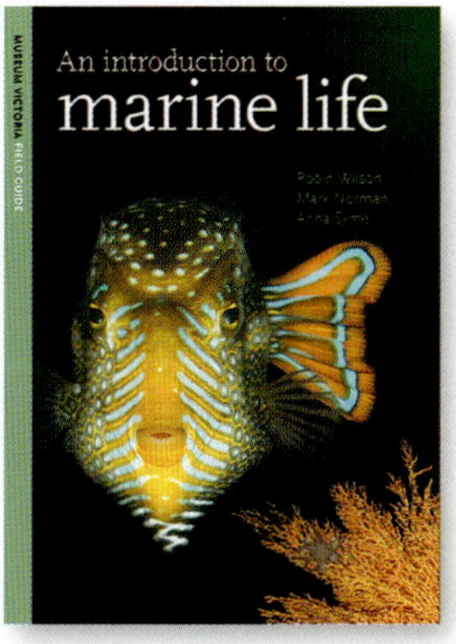

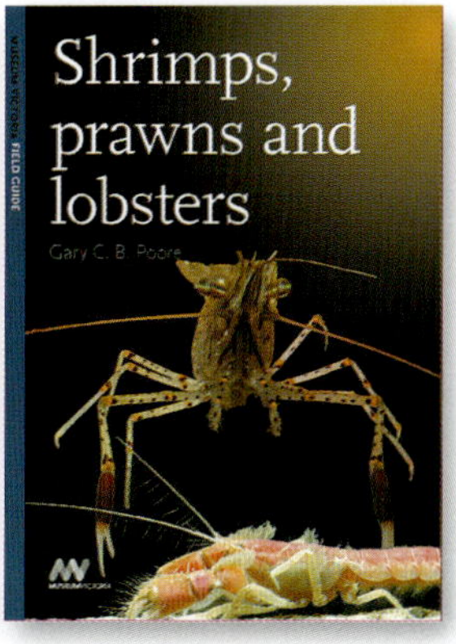